Condensed Matter Physics

Dynamic Correlations

FRONTIERS IN PHYSICS: A Lecture Note and Reprint Series

David Pines, Editor

Volumes of the Series published from 1961 to 1973 are not officially numbered. The parenthetical numbers shown are designed to aid librarians and bibliographers to check the completeness of their holdings.

(1)	N. Bloembergen	Nuclear Magnetic Relaxation: A Reprint Volume, 1961
(2)	G. F. Chew	*S*-Matrix Theory of Strong Interactions: A Lecture Note and Reprint Volume, 1961
(3)	R. P. Feynman	Quantum Electrodynamics: A Lecture Note and Reprint Volume, 1961 (6th printing, 1980)
(4)	R. P. Feynman	The Theory of Fundamental Processes: A Lecture Note Volume, 1961 (5th printing, 1979)
(5)	L. Van Hove, N. M. Hugenholtz, and L. P. Howland	Problems in Quantum Theory of Many-Particle Systems: A Lecture Note and Reprint Volume, 1961
(6)	D. Pines	The Many-Body Problem: A Lecture Note and Reprint Volume, 1961 (5th printing, 1979)
(7)	H. Frauenfelder	The Mössbauer Effect: A Review—with a Collection of Reprints, 1962
(8)	L. P. Kadanoff and G. Baym	Quantum Statistical Mechanics: Green's Function Methods in Equilibrium and Nonequilibrium Problems, 1962 (3rd printing, 1976)
(9)	G. E. Pake	Paramagnetic Resonance: An Introductory Monograph, 1962 [cr. (42)—2nd edition]
(10)	P. W. Anderson	Concepts in Solids: Lectures on the Theory of Solids, 1963 (4th printing, 1978)
(11)	S. C. Frautschi	Regge Poles and *S*-Matrix Theory, 1963
(12)	R. Hofstadter	Electron Scattering and Nuclear and Nucleon Structure: A Collection of Reprints with an Introduction, 1963
(13)	A. M. Lane	Nuclear Theory: Pairing Force Correlations to Collective Motion, 1964
(14)	R. Omnès and M. Froissart	Mandelstam Theory and Regge Poles: An Introduction for Experimentalists, 1963
(15)	E. J. Squires	Complex Angular Momenta and Particle Physics: A Lecture Note and Reprint Volume, 1963
(16)	H. L. Frisch and J. L. Lebowitz	The Equilibrium Theory of Classical Fluids: A Lecture Note and Reprint Volume, 1964
(17)	M. Gell-Mann and Y. Ne'eman	The Eightfold Way: (A Review—with a Collection of Reprints), 1964
(18)	M. Jacob and G. F. Chew	Strong-Interaction Physics: A Lecture Note Volume, 1964
(19)	P. Nozières	Theory of Interacting Fermi Systems, 1964

FRONTIERS IN PHYSICS: A Lecture Note and Reprint Series

David Pines, Editor *(continued)*

(20)	J. R. Schrieffer	Theory of Superconductivity, 1964 (2nd printing, 1971)
(21)	N. Bloembergen	Nonlinear Optics: A Lecture Note and Reprint Volume, 1965 (3rd printing, with addenda and corrections, 1977)
(22)	R. Brout	Phase Transitions, 1965
(23)	I. M. Khalatnikov	An Introduction to the Theory of Superfluidity, 1965
(24)	P. G. de Gennes	Superconductivity of Metals and Alloys, 1966
(25)	W. A. Harrison	Pseudopotentials in the Theory of Metals, 1966 (2nd printing, 1971)
(26)	V. Barger and D. Cline	Phenomenological Theories of High Energy Scattering: An Experimental Evaluation, 1967
(27)	P. Choquard	The Anharmonic Cystal, 1967
(28)	T. Loucks	Augmented Plane Wave Method: A Guide to Performing Electronic Structure Calculations—A Lecture Note and Reprint Volume, 1967
(29)	Y. Ne'eman	Algebraic Theory of Particle Physics: Hadron Dynamics in Terms of Unitary Spin Currents, 1967
(30)	S. L. Adler and R. F. Dashen	Current Algebras and Applications to Particle Physics, 1968
(31)	A. B. Migdal	Nuclear Theory: The Quasiparticle Method, 1968
(32)	J. J. J. Kokkedee	The Quark Model, 1969
(33)	A. B. Migdal and V. Krainov	Approximation Methods in Quantum Mechanics, 1969
(34)	R. Z. Sagdeev and A. A. Galeev	Nonlinear Plasma Theory, 1969
(35)	J. Schwinger	Quantum Kinematics and Dynamics, 1970
(36)	R. P. Feynman	Statistical Mechanics: A Set of Lectures, 1972 (5th printing, 1979)
(37)	R. P. Feynman	Photon-Hadron Interactions, 1972
(38)	E. R. Caianiello	Combinatorics and Renormalization in Quantum Field Theory, 1973
(39)	G. B. Field, H. Arp, and J. N. Bahcall	The Redshift Controversy, 1973 (2nd printing, 1976)
(40)	D. Horn and F. Zachariasen	Hadron Physics at Very High Energies, 1973
(41)	S. Ichimaru	Basic Principles of Plasma Physics: A Statistical Approach, 1973
(42)	G. E. Pake and T. L. Estle	The Physical Principles of Electron Paramagnetic Resonance, 2nd Edition, completely revised, enlarged, and reset, 1973 [cf. (9)—1st edition]

FRONTIERS IN PHYSICS: A Lecture Note and Reprint Series

David Pines, Editor (*continued*)

Volumes published from 1974 onward are being numbered as an integral part of the bibliography:

Number		
43	R. C. Davidson	Theory of Nonneutral Plasmas, 1974
44	S. Doniach and E. H. Sondheimer	Green's Functions for Solid State Physicists, 1974
45	P. H. Frampton	Dual Resonance Models, 1974
46	S. K. Ma	Modern Theory of Critical Phenomena, 1976
47	D. Forster	Hydrodynamic Fluctuations, Broken Symmetry, and Correlation Functions, 1975
48	A. B. Migdal	Qualitative Methods in Quantum Theory, 1977
49	S. W. Lovesey	Condensed Matter Physics: Dynamic Correlations, 1980

Other Numbers in preparation

Condensed Matter Physics

Dynamic Correlations

S. W. Lovesey
Rutherford Laboratory
Oxfordshire, England

1980
The Benjamin/Cummings Publishing Company
Advanced Book Program
Reading, Massachusetts

London · Amsterdam · Don Mills, Ontario · Sydney · Tokyo

CODEN: FRPHA

Library of Congress Cataloging in Publication Data

Lovesey, Stephen W
Condensed matter physics.

(Frontiers in physics; v. 49)
Includes index.
1. Solid state physics. 2. Liquids. 3. Statistical physics. I. Title. II. Series.
QC176.L68 530.4'1 79-25794
ISBN 0-8053-6610-5
ISBN 0-8053-6611-3 pbk.

Manufactured in the United States of America

ABCDEFGHIJ-HA-89876543210

CONTENTS

EDITOR'S FOREWORD

The problem of communicating in a coherent fashion the recent developments in the most exciting and active fields of physics seems particularly pressing today. The enormous growth in the number of physicists has tended to make the familiar channels of communication considerably less effective. It has become increasingly difficult for experts in a given field to keep up with the current literature; the novice can only be confused. What is needed is both a consistent account of a field and the presentation of a definite "point of view" concerning it. Formal monographs cannot meet such a need in a rapidly developing field, and, perhaps more important, the review article seems to have fallen into disfavor. Indeed, it would seem that the people most actively engaged in developing a given field are the people least likely to write at length about it.

FRONTIERS IN PHYSICS has been conceived in an effort to improve the situation in several ways. Leading physicists today frequently give a series of lectures, a graduate seminar, or a graduate course in their special fields of interest. Such lectures serve to summarize the present status of a rapidly developing field and may well constitute the only coherent account available at the time. Often, notes on lectures exist (prepared by the lecturer himself, by graduate students, or by postdoctoral fellows) and are distributed in mimeographed form on a limited basis. One of the principal purposes of the FRONTIERS IN PHYSICS Series is to make such notes available to a wider audience of physicists.

It should be emphasized that lecture notes are necessarily rough and informal, both in style and content; and those in the series will prove no exception. This is as it should be. The point of the series is to offer new, rapid, more informal, and, it is hoped, more effective ways for physicists to teach one another. The point is lost if only elegant notes qualify.

The publication of collections of reprints of recent articles in very active fields of physics will improve communication. Such collections are themselves useful to people working in the field. The value of the reprints will, however, be enhanced if the collection is accompanied by an introduction of moderate length which will serve to tie the collection together and, necessarily, constitute a brief survey of the present status of the field. Again, it is appropriate that such an introduction be informal, in keeping with the active character of the field.

The informal monograph, representing an intermediate step between lecture notes and formal monographs, offers an author the opportunity to present his views of a field which has developed to the point where a summation might prove extraordinarily fruitful but a formal monograph might not be feasible or desirable.

Contemporary classics constitute a particularly valuable approach to the teaching and learning of physics today. Here one thinks of fields that lie at the heart of much of present-day research, but whose essentials are by now well understood, such as quantum electrodynamics or magnetic resonance. In such fields some of the best pedagogical material is not readily available, either because it consists of papers long out of print or lectures that have never been published.

Dynamic correlations are a valuable probe of the behavior of condensed matter systems, from both a theoretical and an experimental viewpoint. Dr. Lovesey has made a number of theoretical contributions which have increased our understanding of neutron scattering and other experiments designed to show dynamic correlations in liquids and solids. He is, therefore, especially well qualified to describe recent developments in this important field; it is a pleasure to welcome him as a contributor to this Series.

David Pines

PREFACE

The aim of the book is to describe the dynamic properties of some models of condensed matter physics of current interest in terms of correlation functions that occur in the interpretation of several types of experiment. Hence, the topics encompass both computational methods, and the dynamic properties of various classical and quantal systems close to thermal equilibrium.

Each of the five main chapters deals with a particular computational method, the use of which is then illustrated by application to several different model systems. In consequence, the main chapters are not arranged in order of increasing difficulty of subject matter. In Chaps. II, III, and IV the emphasis is on hydrodynamic phenomena described by continuum (macroscopic) models, and Chaps. V and VI are concerned mainly with properties described in terms of normal modes, or collective excitations, derived from (microscopic) Hamiltonians. In order to partially offset the limitations imposed by space, time, and the author's fields of competence, a special effort has been made to give references to recent specialist review articles and books, although there is no attempt at an exhaustive bibliography of original papers.

This book is an outcome of preparing lecture courses given at Rice University during my visit there for the calendar year 1978 while on leave from the Institut Laue-Langevin. My visit to Rice University was kindly initiated by Professor G. T. Trammell, and I am very grateful to him and his colleagues in the Physics Department of Rice University for their contributions in making my visit beneficial and pleasurable.

Professor Trammell's assiduity to my lectures, and the frequent classroom discourses he generated together with Professors S. Dodds, H. Huang, and H. Rorschach, and the students led me to subsequently improve some of the material in the original lecture notes. Foibles which remain are entirely my responsibility.

It is not an exaggeration to say that the manuscript would probably not have been finally completed without the considerable secretarial talents of Mary Comerford, whose cheerful countenance was also a fillip to my sagging spirits during some of the more arduous phases of writing.

The task of preparing the manuscript inevitably encroached on time spent with my wife, Margaret, and children, Jane and Kirke. I should like to dedicate the book to them as a token of my gratitude for their unflagging support and forbearance.

ACKNOWLEDGMENTS

Material due to the following authors is reproduced with their kind permission, and that of the The Institute of Physics, London, and *The Physical Review* and *Physical Review Letters:* J. Als-Nielsen, J. F. Cooke, O. W. Dietrich, P. Eisenberger, D. Levesque, J. W. E. Lewis, J. M. Loveluck, A. R. Mackintosh, H. B. Møller, L. Passell, P. M. Platzman, L. Verlet, and C. G. Windsor.

S. W. Lovesey

NOTATION

Because wavevector and frequency dependent quantities occur throughout the book it seemed wise to reserve certain letters to denote wavevectors ($\mathbf{k}$, $\mathbf{p}$, $\mathbf{q}$, and $k = |\mathbf{k}|$, etc.) and frequency (ω). Other letters are reserved for the functions of principal interest, introduced in Chapter 1, namely, the susceptibility (χ), causal Green's function (G), response function (K) and relaxation function (R), and the Hamiltonian ($\mathscr{H}$) and Liouville ($\mathscr{L}$) operators.

The notation $\langle \ldots \rangle$ always denotes the average of the enclosed quantity at a temperature $T = 1/k_B\beta$ where k_B is Boltzmann's constant. Planck's constant is set equal to unity in most parts of the book. In general, an effort has been made to use notation in common practice.

CHAPTER I

INTRODUCTION

1.1 PROLOGUE

This book describes, in terms of correlation functions, the dynamic properties of several classical and quantal model systems of current interest in condensed matter physics. Hence, it is intended primarily for the student seeking an introduction to some theoretical methods of statistical physics that have proved to be very useful and who wants, at the same time, to understand the basic dynamic properties of various model systems. For example, the degenerate electron gas, a good model for understanding electron motion in simple metals and doped semiconductors, has been studied by a large number of authors using a variety of methods. Here we study the degenerate electron gas in zero field by one method, and a second, complementary, method is used to study the striking effects on electron motion produced by a steady magnetic field. As is often the case, the study of similar problems using two methods leads to a better understanding of their strengths and shortcomings.

The discussion of each system, or method, is more or less self-contained so that a reader wishing to study just one particular topic should not find it necessary to refer extensively to many parts of the book. This feature of the presentation is intended to make the book useful to research workers who require, for example, supplementary material to understand an original paper or a basis for interpreting experimental data. In keeping with this, an effort has been made to reference recent specialist review articles and books.

All the models describe systems that are close to thermal equilibrium, and therefore exclude problems like the turbulent motion of fluids [Swinney and Gollub (1978)] and chemical kinetics [Nicolis (1979)]. The models are broadly of two types. For one type of model, the times and lengths of

Stephen Lovesey, Condensed Matter Physics: Dynamic Correlations

ISBN 0-8053-6610-5, 0-8053-6611-3 (pbk.)

interest are long compared to the mean-free time and mean-free path (although these scales are difficult to define precisely except in a dilute gas). In consequence, details of the interatomic potential do not enter the equations-of-motion, and a study of the dynamic properties is based on the equations of continuum mechanics. In the second type of problem the times of interest are comparable to interatomic collision times and the lengths are of the order of interatomic spacings. Hence, in this instance, the dynamic properties of the model are derived from equations-of-motion using a Hamiltonian which describes the model at an atomic level.

The choice of the theoretical method used to derive the dynamic properties depends on the nature of the physical state described by the model. There is a very useful theoretical method which can be used, in certain circumstances, to describe dynamic properties on microscopic and macroscopic scales of time and length, and this method is discussed at length in Chap. III.

While the type of theoretical method used depends to a large extent on the time and length scales of principal interest, the dynamic properties of a model are usually couched in terms of correlation functions, since these are the appropriate functions for the interpretation of data from a variety of important experimental techniques used to study condensed matter. These techniques include resonance methods (nuclear magnetic and electron paramagnetic resonance, Mössbauer spectroscopy) radiation scattering (acoustic waves, and neutron, electron and light scattering) and computer simulation "experiments."

In the remaining sections of this chapter we define a correlation function, and also define some important related functions, such as the dynamic, or generalized, susceptibility. The properties of these functions are illustrated by calculating them for several simple models, including, for example, the harmonic oscillator. We shall motivate the introduction of correlation functions by discussing the form of the scattering cross section for processes that involve the single scattering of radiation. This particular example is chosen for the purpose because the single scattering of radiation, which is describable within the Born approximation, is a very important experimental technique, and the physical process is easily visualized.

The reader should require only a modest background in mathematics and statistical physics to follow the developments in this chapter. The book by Reed and Roy (1971) provides a useful and very readable undergraduate level introduction to statistical physics, and the elementary parts of Schiff's (1955) and Ziman's (1969) texts provide the necessary background in quantum mechanics. For the more advanced topics we shall make frequent references to the textbooks by Landau and Lifshitz, particularly Vol. 5, *Statistical Physics* (1959). An introduction to the theory of thermal neutron scattering from solids and liquids is given by Squires (1978).

ISBN 0-8053-6610-5, 0-8053-6611-3 (pbk.)

1.2 CORRELATION FUNCTIONS

Consider a scattering process in which the state of the target system changes from an initial state described by the wave function $|\lambda\rangle$ to a state $|\lambda'\rangle$. In this process the energy of the target changes from a value E_λ to $E_{\lambda'}$, and conservation of energy requires that the energy of the radiation changes by an amount $\omega = E_{\lambda'} - E_\lambda$. Let us denote the operator which affects this change of state of the target by A. Fermi's Golden Rule, which tells us that the probability for the change to occur, including conservation of energy, is proportional to [see, for example, Schiff (1955) or Landau and Lifshitz (1965)]

$$|\langle\lambda'|A|\lambda\rangle|^2\delta(\omega + E_\lambda - E_{\lambda'}).$$

The total probability is obtained from this result by weighting it with the probability for occupation of the initial state, denoted by p_λ, and summing over all initial and final states. We are therefore led to introduce the function

$$S(\omega) = \sum_{\lambda\lambda'} p_\lambda|\langle\lambda'|A|\lambda\rangle|^2\delta(\omega + E_\lambda - E_{\lambda'}) \tag{1.1}$$

into the interpretation of scattering experiments; $S(\omega)$ is often called the scattering function. A fundamental result of statistical mechanics tells us that the probability of finding a state with energy E_λ is proportional to $\exp(-E_\lambda/k_BT)$ where T is the temperature and k_B is Boltzmann's constant, and $p_\lambda = \exp(-E_\lambda/k_BT)/Z$ where the partition function Z makes $\sum p_\lambda = 1$.

$S(\omega)$ is expressed in terms of a correlation function by using the well-known representation of the delta function

$$\delta(\omega) = \frac{1}{2\pi}\int_{-\infty}^{\infty} dt \exp(-i\omega t) \tag{1.2}$$

in which the integration variable t has the dimension of time (Planck's constant is set equal to unity in most parts of the book). Using (1.2) in (1.1), we obtain

$$S(\omega) = \frac{1}{2\pi}\int_{-\infty}^{\infty} dt \exp(-i\omega t) \sum_{\lambda\lambda'} p_\lambda\langle\lambda|A^+|\lambda'\rangle\langle\lambda'| \exp(itE_{\lambda'}) \cdot A \exp(-itE_\lambda)|\lambda\rangle \tag{1.3}$$

where we have also used the fact that the complex conjugate of a matrix element can be written in terms of the matrix element of the Hermitian conjugate of the operator, namely,

$$\langle\lambda'|A|\lambda\rangle^* = \langle\lambda|A^+|\lambda'\rangle. \tag{1.4}$$

ISBN 0-8053-6610-5, 0-8053-6611-3 (pbk.)

Next, we define a Heisenberg operator

$$A(t) = \exp(it\mathcal{H})A\exp(-it\mathcal{H}) \tag{1.5}$$

where $\mathcal{H}$ is the Hamiltonian that describes the target system. Clearly,

$$\langle\lambda'|\exp(itE_{\lambda'})A\exp(-itE_{\lambda})|\lambda\rangle = \langle\lambda'|A(t)|\lambda\rangle. \tag{1.6}$$

When we insert this result into (1.3) the sum over the states $|\lambda'\rangle$ can be performed by closure, i.e.

$$\sum_{\lambda'} |\lambda'\rangle\langle\lambda'| = 1. \tag{1.7}$$

The autocorrelation function of the variable A is defined to be ($A(0) \equiv A$),

$$\begin{aligned}\langle A^+A(t)\rangle &= \sum_{\lambda} p_\lambda\langle\lambda|A^+A(t)|\lambda\rangle \\ &= Z^{-1}\,\mathrm{Tr}\exp(-\mathcal{H}/k_BT)A^+A(t)\end{aligned} \tag{1.8}$$

where Tr denotes the trace operation. In most cases $\langle A\rangle = 0$.

The desired form for the scattering function as the Fourier transform of an autocorrelation function is then [Van Hove (1954)]

$$S(\omega) = \frac{1}{2\pi}\int_{-\infty}^{\infty} dt\exp(-i\omega t)\langle A^+A(t)\rangle. \tag{1.9}$$

In this form it is clear that $S(\omega)$ is the spectrum of *spontaneous* fluctuations in A.

Thus far we have omitted all reference to the change in the wavevector of the radiation that takes place in the scattering experiment. The wavevector dependence of the scattering function is hidden in the operator A. In the Born approximation for scattering, the cross section is given by the expression (1.1) and A is the operator formed by taking matrix elements of the interaction operator between initial and final plane wave states [Schiff (1955) or Landau and Lifshitz (1965).]. Thus, if the initial and final wavevectors of the scattered radiation are $\mathbf{k}$ and $\mathbf{k}'$, respectively, and the interaction operator is $V(\mathbf{r})$, then

$$A = \int d\mathbf{r}\exp(-i\mathbf{k}'\cdot\mathbf{r})V(\mathbf{r})\exp(i\mathbf{k}\cdot\mathbf{r}). \tag{1.10}$$

In most cases of interest, the dependence of the matrix element on $\mathbf{k}$ and $\mathbf{k}'$ involves only the scattering vector $\mathbf{Q} = \mathbf{k} - \mathbf{k}'$, and A is then the spatial Fourier transform, $A(\mathbf{Q})$, of the interaction operator. A always takes this simple form, $A(\mathbf{Q})$, when although the Born approximation is not adequate for the true interaction operator, it becomes so upon replacing the true interaction by a suitable pseudo-potential as, for example, in the nonresonant nuclear scattering of slow neutrons [Squires (1978)]. Because $V(\mathbf{r})$ is

ISBN 0-8053-6610-5, 0-8053-6611-3 (pbk.)

Hermitian,

$$\begin{aligned}\langle\lambda'|A(\mathbf{Q})|\lambda\rangle^* &= \int d\mathbf{r}\,\exp(-i\mathbf{Q}\cdot\mathbf{r})\langle\lambda'|V(\mathbf{r})|\lambda\rangle^* \\ &= \int d\mathbf{r}\,\exp(-i\mathbf{Q}\cdot\mathbf{r})\langle\lambda|V(\mathbf{r})|\lambda'\rangle \\ &= \langle\lambda|A(-\mathbf{Q})|\lambda'\rangle.\end{aligned}$$

In view of this result, the full frequency and wavevector dependent scattering function is obtained from (1.9) making the replacements $A^+ \to A(-\mathbf{Q}, 0)$ and $A(t) \to A(\mathbf{Q}, t)$.

Nothing that we have discussed so far is materially altered by including the wavevector dependence of the scattering process and it is tempting, therefore, to continue keeping our notation simple and to omit explicit reference to the wavevector. However, we shall have occasion in subsequent discussions to use the fact that $V(\mathbf{r})$ is Hermitian and our bookkeeping will break down if we take A in (1.9) to be Hermitian, i.e. if we delete the superscript + on the left-hand operator in the correlation function. We choose to continue with the notation adopted in (1.9). If the wavevector dependence of the problem is to be omitted then $A^+ = A$, while if the wavevector dependence is retained $A^+ = A(-\mathbf{Q})$.

Before discussing some important general features of autocorrelation functions, we examine the form of $S(\omega)$ for three simple models of interest.

1.3 EXAMPLES

As a first example we shall calculate the displacement autocorrelation function for a single particle bound in an isotropic, harmonic oscillator potential. This correlation function is of great importance in the theory of scattering from materials since it governs the lowest-order inelastic process. We shall return to a discussion of this point at the end of the section.

Let us consider the displacement in one direction in the crystal, and denote the corresponding displacement by u. For a harmonic theory, the Hamiltonian of the single atom in question is

$$\mathcal{H} = \omega_0 a^+ a \tag{1.11}$$

where a and a^+ are the usual Bose operators, ω_0 is the frequency of vibration, and the displacement of a particle of unit mass is

$$u = (2\omega_0)^{-1/2}(a + a^+). \tag{1.12}$$

The Heisenberg operator $a(t)$ is found from the equation-of-motion,*

$$i\partial_t a(t) = [a(t), \mathcal{H}], \tag{1.13}$$

* We shall often use the notation ∂_t ($\equiv \partial/\partial t$) for partial derivatives.

ISBN 0-8053-6610-5, 0-8053-6611-3 (pbk.)

where [,] is a commutator, with the result ($a(0) \equiv a$),

$$a(t) = \exp(-i\omega_0 t)a. \tag{1.14}$$

With these results,

$$\begin{aligned}\langle A^+A(t)\rangle &= \langle uu(t)\rangle \\ &= (2\omega_0)^{-1}\{\langle aa^+\rangle \exp(i\omega_0 t) + \langle a^+a\rangle \exp(-i\omega_0 t)\} \\ &= (2\omega_0)^{-1}\{(1 + n_0)\exp(i\omega_0 t) + n_0 \exp(-i\omega_0 t)\},\end{aligned} \tag{1.15}$$

where n_0 is the Bose distribution function for a temperature $T = 1/k_B\beta$,

$$n_0 = \{\exp(\beta\omega_0) - 1\}^{-1}. \tag{1.16}$$

Using (1.15) in (1.9) leads to the desired result for the scattering function

$$S(\omega) = (2\omega_0)^{-1}\{(1 + n_0)\delta(\omega - \omega_0) + n_0\delta(\omega + \omega_0)\}. \tag{1.17}$$

The two terms in (1.17) can be viewed as representing the creation and annihilation of a quanta of frequency ω_0. In the limit of low temperatures, where $\beta\omega_0 \to \infty$, $n_0 \sim 0$, and, in consequence, only the creation process occurs with an appreciable weight since there is a negligible fraction of thermally excited quanta that can be annihilated.

As a second example we consider scattering from a single atom of mass M whose equilibrium state is described by a Boltzmann distribution. For this case the variable A is the spatial Fourier transform of the particle density, namely

$$A(t) = \int d\mathbf{r} \exp(i\mathbf{Q}\cdot\mathbf{r})\delta(\mathbf{r} - \mathbf{R}(t)) = \exp(i\mathbf{Q}\cdot\mathbf{R}(t)), \tag{1.18}$$

where $\mathbf{R}$ denotes the position of the atom, and $\mathbf{Q}$ is the change in the wavevector of the neutrons or X-rays. An expression for $A(t)$ is obtained from its equation-of-motion, cf. Eq. (1.13), using $\mathcal{H} = \mathbf{p}^2/2M$ where $\mathbf{p}$ is the momentum conjugate to $\mathbf{R}$,

$$\begin{aligned}i\partial_t A(t) &= \frac{1}{2M}[\exp(i\mathbf{Q}\cdot\mathbf{R}(t)), p^2] \\ &= \frac{1}{2M}\exp(i\mathcal{H}t)[\exp(i\mathbf{Q}\cdot\mathbf{R}), p^2]\exp(-i\mathcal{H}t) \\ &= -\frac{1}{2M}A(t)(2\mathbf{Q}\cdot\mathbf{p} + Q^2)\end{aligned} \tag{1.19}$$

where we have used the identity $[A, p^2] = [A, \mathbf{p}]\cdot\mathbf{p} + \mathbf{p}\cdot[A, \mathbf{p}]$ and the commutation relation $[A, \mathbf{p}] = -\mathbf{Q}A$. From this equation we find immediately,

$$A(t) = \exp(i\mathbf{Q}\cdot\mathbf{R})\exp\{(it/2M)(2\mathbf{Q}\cdot\mathbf{p} + Q^2)\} \tag{1.20}$$

ISBN 0-8053-6610-5, 0-8053-6611-3 (pbk.)

and so ($\beta = 1/k_B T$)

$$\langle A^+ A(t)\rangle = \exp(itQ^2/2M)\langle\exp(it\mathbf{Q}\cdot\mathbf{p}/M)\rangle$$
$$= \exp(itQ^2/2M)\int d\mathbf{p}$$
$$\cdot \exp\left(-\frac{\beta p^2}{2M} + \frac{it\mathbf{Q}\cdot\mathbf{p}}{M}\right)\Big/\int d\mathbf{p}\exp\left(-\frac{\beta p^2}{2M}\right)$$
$$= \exp\{-(Q^2/2M)(t^2/\beta - it)\}. \tag{1.21}$$

The corresponding scattering function is,

$$S(Q, \omega) = \left(\frac{M\beta}{2\pi Q^2}\right)^{1/2} \exp\left\{-\frac{M\beta}{2Q^2}\left(\omega - \frac{Q^2}{2M}\right)^2\right\}. \tag{1.22}$$

The scattered radiation spectrum is seen to be centered about an energy transfer $\omega = Q^2/2M$, and to have an energy width $\sim(Q^2/M\beta)^{1/2}$ which increases with Q and temperature. The limit $M \to \infty$ of (1.22) corresponds to scattering from a single fixed atom, and the limit can be affected with the aid of the result

$$\delta(b) = \lim_{\epsilon\to 0}(\pi\epsilon)^{-1/2}\exp(-b^2/\epsilon). \tag{1.23}$$

We consider next the corresponding calculation for a single atom bound in a harmonic potential. The contribution to the scattering function associated with a single quanta will be proportional to (1.17) for the displacement autocorrelation function.

For an isotropic harmonic potential we need only consider the motion in one direction in space. In (1.18) we denote the displacement of the atom from its equilibrium position by $u(t)$, then the correlation function of interest is

$$\langle\exp(-iQu)\exp(iQu(t))\rangle. \tag{1.24}$$

From (1.12) and (1.14) we find

$$u(t) = (2\omega_0)^{-1/2}\{a\exp(-\omega_0 t) + a^+\exp(i\omega_0 t)\}. \tag{1.25}$$

Because the commutator $[u(t), u]$ is a c-number we can use the following identity to combine the exponentials in (1.24),*

$$\exp A\cdot\exp B = \exp(A + B + \tfrac{1}{2}[A, B]). \tag{1.26}$$

Thus,

$$\langle\exp(-iQu)\exp(iQu(t))\rangle$$
$$= \exp\{\tfrac{1}{2}Q^2[u, u(t)]\}\langle\exp\{iQ[u(t) - u]\}\rangle$$
$$= \exp\{\tfrac{1}{2}Q^2[u, u(t)]\}\exp\{-\tfrac{1}{2}Q^2\langle(u(t) - u)^2\rangle\}$$
$$= \exp\{-Q^2(\langle u^2\rangle - \langle uu(t)\rangle)\}. \tag{1.27}$$

* For a detailed proof see, for example, Englman and Levy (1963).

ISBN 0-8053-6610-5, 0-8053-6611-3 (pbk.)

In obtaining the second equality we have used the following identity for an arbitrary linear combination of Bose operators A,

$$\langle \exp A \rangle = \exp(\tfrac{1}{2}\langle A^2 \rangle). \tag{1.28}$$

A proof of (1.28) is given in Appendix 1A. The correlation functions required in (1.27) are obtained from (1.15), and the result is

$$\begin{aligned}&\langle \exp(-iQu)\exp(iQu(t))\rangle \\ &\quad = \exp(-2W)\exp\left\{\frac{Q^2}{4\omega_0}\operatorname{cosech}(\tfrac{1}{2}\omega_0\beta)\right. \\ &\qquad \left.\times\,[\exp(i\omega_0 t + \tfrac{1}{2}\omega_0\beta) + \exp(-i\omega_0 t - \tfrac{1}{2}\omega_0\beta)]\right\}. \end{aligned} \tag{1.29}$$

The first factor in (1.29) is usually called the Debye-Waller factor [Squires (1978)], and W is given explicitly by

$$W = \tfrac{1}{2}Q^2\langle u^2\rangle = Q^2\frac{1}{4\omega_0}\coth(\tfrac{1}{2}\beta\omega_0). \tag{1.30}$$

The remaining factor in (1.29) can be put into a more compact form by using the identity

$$\exp\{\tfrac{1}{2}y(t + 1/t)\} = \sum_{n=-\infty}^{\infty} t^n I_n(y), \qquad n = 0, \pm 1, \pm 2, \ldots \tag{1.31}$$

where $I_n(y)$ is the modified Bessel function of the first kind, and $I_n(y) = I_{-n}(y)$. With

$$y = Q^2\frac{1}{2\omega_0}\operatorname{cosech}(\tfrac{1}{2}\omega_0\beta)$$

we find, finally,

$$\begin{aligned}&\langle \exp(-iQu)\exp(iQu(t))\rangle \\ &\qquad = \exp(-2W)\sum_{n=-\infty}^{\infty} I_n(y)\exp(\tfrac{1}{2}\omega_0\beta n + i\omega_0 t n)\end{aligned} \tag{1.32}$$

and

$$\begin{aligned}S(Q,\omega) &= \frac{1}{2\pi}\int_{-\infty}^{\infty} dt\,\exp(-i\omega t)\langle \exp(-iQu)\exp(iQu(t))\rangle \\ &= \exp(-2W + \tfrac{1}{2}\omega\beta)\sum_{n=-\infty}^{\infty} I_n(y)\delta(\omega - n\omega_0).\end{aligned} \tag{1.33}$$

This last result shows clearly that the integer n measures the number of quanta involved in the nth term in the series.

The term $n = 0$ corresponds to an elastic process in which no energy

ISBN 0-8053-6610-5, 0-8053-6611-3 (pbk.)

is exchanged between the scattered radiation and the target. The terms with $n = 1$ and $n = -1$ are the single quanta contributions to the scattering function. The contribution that corresponds to (1.17) is obtained from (1.33) by just keeping the terms $n = \pm 1$ and expanding $I_1(y)$ and the Debye-Waller factor in Q^2. By doing so we recover (1.17) apart from an additional factor of Q^2. The reader might like to check this by expanding the exponentials in the correlation function (1.24) to first-order in Q and using (1.15).

We shall now consider some general properties of autocorrelation and scattering functions needed in the discussion of linear response theory given in Sec. 1.5.

1.4 SOME GENERAL RESULTS

If the model system is stationary, as we shall always assume, then by definition

$$\langle A^+(t_0)A(t + t_0)\rangle = \langle A^+A(t)\rangle, \tag{1.34}$$

where the time t_0 is arbitrary. Taking $t_0 = -t$ we obtain the useful result,

$$\langle A^+(-t)A\rangle = \langle A^+A(t)\rangle, \tag{1.35}$$

where $A \equiv A(0)$. Using (1.4) it is easy to verify the following important identity for the complex conjugate of a correlation function,

$$\langle A^+A(t)\rangle^* = \langle A^+(t)A\rangle. \tag{1.36}$$

The results (1.35) and (1.36) enable us to prove that $S(\omega)$ is a real function as, of course, it should be since it is the spectral distribution for a scattering process.

Notice that in general an autocorrelation function is a complex quantity. In fact, it is only for systems which obey the laws of classical statistical mechanics that correlation functions are always purely real. This point is nicely illustrated with the result (1.21); when we insert Planck's constant $\hbar$ we have

$$\langle\exp\{-i\mathbf{Q}\cdot\mathbf{R}\}\exp\{i\mathbf{Q}\cdot\mathbf{R}(t)\}\rangle = \exp\{-(Q^2/2M)(t^2/\beta - i\hbar t)\}. \tag{1.21a}$$

The right-hand side of this expression is purely real in the classical limit $\hbar \to 0$, and the corresponding scattering function is a Gaussian centered at $\omega = 0$. Looking back at the derivation of (1.21) it is seen that the occurrence of the term in (1.21a) that contains $\hbar$ is a direct consequence of the fact that the operators $\mathbf{p}$ and $\exp(i\mathbf{Q}\cdot\mathbf{R})$ do not commute.

It is shown in Appendix 1.B that

$$\langle A^+A(t)\rangle = \langle A(t - i\hbar\beta)A^+\rangle. \tag{1.37}$$

From this result we deduce that in the classical limit $\hbar \to 0$ a correlation function is an even function of time. Moreover, an autocorrelation function for a classical system is purely real, as we have just remarked. We now

ISBN 0-8053-6610-5, 0-8053-6611-3 (pbk.)

return to the convention $\hbar = 1$. If we use (1.37) in conjunction with (1.35) we find

$$S(\omega) = S(-\omega)\exp(\beta\omega). \tag{1.38}$$

This relation is usually referred to as the condition of detailed balance. It is worthwhile to verify that the results (1.17), (1.22), and (1.33) satisfy the condition (1.38). Strictly (1.38) should read

$$S(\mathbf{Q}, \omega) = S(-\mathbf{Q}, -\omega)\exp(\beta\omega), \tag{1.38a}$$

but if the target has inversion symmetry then $S(\mathbf{Q}, \omega) = S(-\mathbf{Q}, \omega)$, and (1.38) with $\mathbf{Q}$ implicit in the scattering function is valid.

If (1.34) is differentiated with respect to t_0 then

$$\langle\{\partial_{t_0}A^+(t_0)\}A(t + t_0)\rangle + \langle A^+(t_0)\{\partial_{t_0}A(t + t_0)\}\rangle = 0.$$

Employing the frequently used notation $\partial_t A = \dot{A}$, and setting $t_0 = 0$ we find

$$\langle\dot{A}^+A(t)\rangle = -\langle A^+\dot{A}(t)\rangle. \tag{1.39}$$

The identity (1.39) can also be derived by using the equation-of-motion for A, namely $\dot{A} = -i[A, \mathcal{H}]$, and the invariance of the trace operation in (1.8) to a cyclic permutation of the operators.

1.5 LINEAR RESPONSE THEORY

An important approach to the study of the nonequilibrium properties of a particular system is to calculate the response of the system to an applied time-dependent perturbation. If, for example, the system supports collective excitations, e.g. phonons, and the applied perturbation couples to the excitations then the measured response will be large. Usually, we measure the response of a system to a Fourier component of the applied time-dependent perturbation in terms of a dynamic susceptibility. There is a very direct relation between the spectrum of spontaneous fluctuations, described by the function $S(\omega)$, and the dynamic susceptibility called the fluctuation-dissipation theorem [Kubo (1966)], and this is derived after the following calculation of the linear response to an applied perturbation.

We calculate the response of a system to a time-dependent perturbation by calculating, to first order, the change induced in the density matrix [Ziman (1969)]. At the time $t = -\infty$ the system is described by a Hamiltonian $\mathcal{H}_0$ and at some later time the total Hamiltonian is taken to be

$$\mathcal{H} = \mathcal{H}_0 - A^+h(t). \tag{1.40}$$

The scalar function $h(t)$ measures the strength of the perturbation whose coupling to the system is described by the operator A^+. The equation-of-

ISBN 0-8053-6610-5, 0-8053-6611-3 (pbk.)

motion of the density matrix $\mu(t)$ is

$$i\partial_t\mu(t) = -[\mu(t), \mathscr{H}(t)]$$

and, $\langle A(t)\rangle = \mathrm{Tr}\ \mu(t)A$.

We write $\mu(t) = \mu_0 + \delta\mu(t)$ where μ_0 describes the system at $t = -\infty$ when $h = 0$. Neglecting a second-order term, the *linear* equation for $\delta\mu(t)$ is

$$i\partial_t\delta\mu(t) = [\mathscr{H}_0, \delta\mu(t)] + h(t)[\mu_0, A^+].$$

It is convenient to introduce an auxiliary function $y(t)$ defined by

$$\delta\mu(t) = \exp(-i\mathscr{H}_0 t)y(t)\exp(i\mathscr{H}_0 t)$$

since then

$$i\partial_t\delta\mu(t) = [\mathscr{H}_0, \delta\mu(t)] + \exp(-i\mathscr{H}_0 t)\{i\partial_t y(t)\}\exp(i\mathscr{H}_0 t)$$

and

$$i\partial_t y(t) = h(t)[\mu_0, A^+(t)].$$

From this last result we readily obtain the desired expression for $\delta\mu(t)$,

$$i\delta\mu(t) = \int_{-\infty}^{t} d\bar{t}\ h(\bar{t})[\mu_0, A^+(\bar{t} - t)]. \tag{1.41}$$

The change in the variable A is then

$$\mathrm{Tr}\ \delta\mu(t)A = -i\int_{-\infty}^{t} d\bar{t}\ h(\bar{t})\ \mathrm{Tr}\{[\mu_0, A^+(\bar{t} - t)]A\}$$

$$= \int_{-\infty}^{t} d\bar{t}\ h(\bar{t})K(t - \bar{t}). \tag{1.42}$$

where the second equality defines the linear response function $K(t)$. Using the stationary condition (1.35) and the cyclic properties of the trace operation $K(t)$ can be expressed in the convenient form,

$$K(t) = i\,\mathrm{Tr}\{\mu_0[A(t), A^+]\}$$

$$= i\langle[A(t), A^+]\rangle. \tag{1.43}$$

We pause at this juncture to discuss some features of the important result (1.42), and our definition of the response function. Because a response is detected only after (or possibly at the instant) the perturbation is applied, the range of integration in (1.42) is over all times prior to the observation time, t. In consequence, the time argument of the response function is positive under the integral. This feature of (1.42), and the associated response function, is usually called the causal property of the response.

ISBN 0-8053-6610-5, 0-8053-6611-3 (pbk.)

It can be rightly argued that causality requires that a response function can be defined only in the context of (1.42), where its position in the integral guarantees that its time argument is positive. Alternatively, if a response function is defined in isolation from (1.42) then it would seem that the definition should be such that the function is zero for negative time arguments, and this can be achieved by including a unit step function in the definition. In fact, we shall introduce such a function in Sec. 1.6, where it is called a causal Green's function. It is, however, convenient in the mathematical development that we choose to pursue to have a function that is defined for positive and negative times by the commutator in (1.43), and we shall refer to this function, $K(t)$, as the response function. Later in this chapter we consider how the response to a discontinuous perturbation decays in time after the perturbation has been switched off. The physics of the situation again requires that the appropriate function, called the relaxation function, exists only for positive time arguments. But here also it is convenient in the mathematical development to define the relaxation function to be finite for both positive and negative time arguments.

The identities (1.35) and (1.37) enable us to prove the following important relation between the response function and the scattering function,

$$\{1 - \exp(-\beta\omega)\}S(\omega) = \frac{1}{2\pi i}\int_{-\infty}^{\infty} dt\, K(t)\exp(i\omega t). \tag{1.44}$$

This relation, often called the fluctuation-dissipation theorem, relates the spontaneous fluctuation properties of a system in equilibrium, described by $S(\omega)$, to the response $K(t)$ of the system to an external time dependent perturbation.

Using the identity (1.36) we can show that $K(t)$ is purely real, and in doing so it is best to display explicitly the wavevector dependence. From (1.36) and (1.43),

$$\begin{aligned} K^*(Q, t) &= -i\langle A(\mathbf{Q}, t)A(-\mathbf{Q}, 0) - A(-\mathbf{Q}, 0)A(\mathbf{Q}, t)\rangle^* \\ &= -i\langle A^+(-\mathbf{Q}, 0)A^+(\mathbf{Q}, t) - A^+(\mathbf{Q}, t)A^+(-\mathbf{Q}, 0)\rangle \\ &= i\langle A(-\mathbf{Q}, t)A(\mathbf{Q}, 0) - A(\mathbf{Q}, 0)A(-\mathbf{Q}, t)\rangle \\ &= K(-Q, t) = K(Q, t), \end{aligned}$$

where in the last equality we assume that the system has inversion symmetry. Because $S(Q, \omega)$ is also purely real we deduce from (1.44) that $K(Q, t)$ is an odd function of time, i.e.

$$K(Q, t) = -K^*(Q, -t) = -K(Q, -t). \tag{1.45}$$

We now wish to define a dynamic susceptibility and a convenient procedure is to choose a particular representation for the strength function

ISBN 0-8053-6610-5, 0-8053-6611-3 (pbk.)

in (1.40). With the choice

$$h(t) = h_0 \exp(i\omega t + \eta t) \tag{1.46}$$

and $\eta > 0$ we satisfy the requirement $h(-\infty) = 0$, and the change in the variable A is

$$\mathrm{Tr}\, \delta\mu(t)A = h(t) \int_0^\infty d\bar{t}\, K(\bar{t}) \exp(-i\omega\bar{t} - \eta\bar{t}). \tag{1.47}$$

If we define a dynamic susceptibility $\tilde{\chi}$ as the Laplace transform of $K(t)$,

$$\tilde{\chi}(s) = \int_0^\infty dt\, K(t) \exp(-st), \tag{1.48}$$

then we obtain from (1.47) a simple relation between the change in A induced by the perturbation described by (1.46) and the dynamic susceptibility, namely,

$$\mathrm{Tr}\, \delta\mu(t)A = h(t)\tilde{\chi}(i\omega + \eta). \tag{1.49}$$

Landau and Lifshitz (1959) refer to $\tilde{\chi}(-i\omega)$ as the generalized susceptibility, and this terminology is used by many other authors.

Since $S(\omega)$ is proportional to the Fourier transform of $K(t)$ and $\tilde{\chi}(s)$ is defined as the Laplace transform of $K(t)$ there is evidently a simple relation between $S(\omega)$ and $\tilde{\chi}(i\omega)$. On using the fact that $K(t)$ is an odd function of t, we find

$$\{1 - \exp(-\beta\omega)\}S(\omega) = \frac{1}{\pi}\tilde{\chi}''(-i\omega) = \frac{-1}{\pi}\tilde{\chi}''(i\omega), \tag{1.50}$$

where $\tilde{\chi}''(i\omega)$ is the imaginary part of the dynamic susceptibility.

1.6 RELAXATION AND CAUSAL GREEN'S FUNCTIONS

Instead of calculating the dynamic susceptibility from (1.49) it is sometimes more convenient to work with closely related functions, two of which we now define and discuss.

The relaxation function describes the behavior of the change in a variable after the external perturbation has been switched off. An explicit expression for the relaxation function is obtained from (1.42) by taking $h(t)$ to be $h_0 \exp(\eta t)$ for $t \leq 0$ and zero for $t > 0$. The change in A induced by this discontinuous perturbation is then

$$\mathrm{Tr}\delta\mu(t)A = \int_{-\infty}^0 d\bar{t}\, h_0 \exp(\eta\bar{t}) K(t - \bar{t})$$

$$= \int_t^\infty d\bar{t}\, h_0 \exp\{\eta(t - \bar{t})\} K(\bar{t}).$$

ISBN 0-8053-6610-5, 0-8053-6611-3 (pbk.)

In view of this result we define, following Kubo (1966), the relaxation function $R(t)$ as,

$$R(t) = \int_t^\infty d\bar{t}\, K(\bar{t}) \tag{1.51a}$$

or

$$\partial_t R(t) = -K(t). \tag{1.51b}$$

We shall use (1.51b) to define $R(t)$ for positive and negative time arguments. The change in A induced by a discontinuous perturbation is expected to vanish after a long time from the time of switching off and, in consequence, $R(t = \infty) = 0$ gives the constant of integration in (1.51a). Since $K(t)$ is an odd function of t, $R(t)$ is an even function of time. From (1.48) and (1.51b) it follows that

$$-R(0) + s\tilde{R}(s) = -\tilde{\chi}(s), \tag{1.52}$$

and setting $s = i\omega$, and assuming $R(0)$ to be real, we find

$$\tilde{\chi}''(i\omega) = -\omega\tilde{R}'(i\omega) \tag{1.53}$$

where $\tilde{R}'(i\omega)$ is the real part of the Laplace transform of $R(t)$.

In constructing an explicit form for $R(t)$ we shall recall that the probability function p_λ in (1.8) is

$$p_\lambda = Z^{-1}\exp(-\beta E_\lambda)$$

where the partition function Z is defined by

$$Z = \sum_\lambda \exp(-\beta E_\lambda)$$

so that $\Sigma\, p_\lambda = 1$. Starting from (1.43) and going to (1.51) it is shown in Appendix 1C that

$$R(t) = \int_0^\beta d\tau\, \langle A^+(-i\tau)A(t)\rangle - \text{constant}.$$

This result is, perhaps, most easily verified by differentiating with respect to t and showing that it satisfies (1.51b);

$$\begin{aligned}
\partial_t R(t) &= \int_0^\beta d\tau\, \langle A^+(-i\tau)\partial_t A(t)\rangle \\
&= -i\int_0^\beta d\tau\, \langle\{\partial_\tau A^+(-i\tau)\}A(t)\rangle \\
&= -i\langle A^+(-i\beta)A(t) - A^+A(t)\rangle \\
&= -i\langle[A(t), A^+]\rangle,
\end{aligned}$$

ISBN 0-8053-6610-5, 0-8053-6611-3 (pbk.)

where to obtain the final result we have used (1.37). The constant of integration in $R(t)$ is chosen so as to make $R(t = \infty) = 0$. The law of increase in entropy, or loss of information, requires that there is no correlation between processes that are well separated in time, and, in consequence,

$$\lim_{t \to \infty} \langle BA(t) \rangle = \langle B \rangle \langle A \rangle, \tag{1.54}$$

where A and B are arbitrary operators. This result would not hold, for example, if there was a persistent mode in the system to which either A or B couple. However, in reality such modes do not exist since there is an infinite number of degrees of freedom and the mode will, eventually, decay.

In view of this discussion we define,

$$R(t) = \int_0^\beta d\tau\, A\langle {}^+(-i\tau) A(t) \rangle - \beta \langle A^+ \rangle \langle A \rangle \tag{1.55}$$

and

$$R(0) = \int_0^\beta d\tau \,\langle A^+(-i\tau) A \rangle - \beta \langle A^+ \rangle \langle A \rangle$$

$$= \int_0^\beta d\tau \,\langle \{A^+(-i\tau) - \langle A^+ \rangle\}\{A - \langle A \rangle\} \rangle = \chi \tag{1.56}$$

where χ is the static susceptibility. It follows from (1.48) and (1.51a) that $\chi = \tilde{\chi}\ (s = 0)$, as expected.

To verify the third equality in (1.56) we modify the Hamiltonian so that

$$\mathcal{H} = \mathcal{H}_0 - \mathcal{H}_1 \tag{1.57}$$

where $\mathcal{H}_1$ is a small time-independent perturbation. The change in the variable A induced by $\mathcal{H}_1$ is

$$\delta\langle A \rangle = Z^{-1} \operatorname{Tr}\{A \exp(-\beta\mathcal{H})\} - Z_0^{-1} \operatorname{Tr}\{A \exp(-\beta\mathcal{H}_0)\}$$

$$= Z^{-1} \operatorname{Tr}\{A \exp(-\beta\mathcal{H})\} - \langle A \rangle \tag{1.58}$$

with the partition function, $Z = \operatorname{Tr} \exp(-\beta\mathcal{H})$. If we define the function $\Phi(\beta)$ through

$$\exp\{-\beta(\mathcal{H}_0 - \mathcal{H}_1)\} = \exp(-\beta\mathcal{H}_0)\, \Phi(\beta)$$

then it satisfies

$$\partial_\beta \Phi(\beta) = \mathcal{H}_1(-i\beta)\, \Phi(\beta) \tag{1.59}$$

with the boundary condition $\Phi(0) = 1$. Iterating this equation once gives,

$$\Phi(\beta) \simeq 1 + \int_0^\beta d\tau\, \mathcal{H}_1(-i\tau), \tag{1.60}$$

$$Z \simeq Z_0(1 + \beta\langle \mathcal{H}_1 \rangle) \tag{1.61}$$

ISBN 0-8053-6610-5, 0-8053-6611-3 (pbk.)

and

$$Z^{-1}\,\mathrm{Tr}\{A\exp(-\beta\mathcal{H})\} \simeq \{\langle A\rangle + \int_0^\beta d\tau\,\langle \mathcal{H}_1(-i\tau)A\rangle\}(1-\beta\langle\mathcal{H}_1\rangle). \quad (1.62)$$

We now choose $\mathcal{H}_1 = hA^+$ and define the static susceptibility,

$$\begin{aligned}\chi &= \delta\langle A\rangle/h \\ &= \int_0^\beta d\tau\,\langle A^+(-i\tau)A\rangle - \beta\langle A^+\rangle\langle A\rangle \end{aligned} \quad (1.63)$$

where the second line follows from (1.58) and (1.62).

The second function of interest in this section is the causal Green's function* defined by

$$\begin{aligned}G(t) &= -i\theta(t)\langle[A(t), A^+]\rangle = -\theta(t)K(t) = \theta(t)\partial_t R(t) \\ &= \langle\langle A(t); A^+\rangle\rangle \end{aligned} \quad (1.64)$$

where $\theta(t)$ is the unit step function ($\theta(t) = 1$ for $t \geq 0$ and zero otherwise) and the fourth equality defines a convenient notation. Notice that the important result (1.42) for the change in the average value of A induced by the perturbation in (1.40) can be written

$$\mathrm{Tr}\,\delta\mu(t)A = -\int_{-\infty}^{\infty} d\bar{t}\, h(t-\bar{t})G(\bar{t}). \quad (1.42\mathrm{a})$$

Let us begin our discussion of $G(t)$ by forming its equation-of-motion. Using the result $\partial_t\theta(t) = \delta(t)$ and the equation-of-motion for $A(t)$ we find

$$i\partial_t G(t) = \delta(t)\langle[A, A^+]\rangle + \langle\langle[A(t), \mathcal{H}]; A^+\rangle\rangle. \quad (1.65)$$

If $h(t)$ in (1.42a) is proportional to $\exp(i\omega t)$, as in (1.46), then it is convenient to define the Fourier transform of $G(t)$,

$$\begin{aligned}G(\omega) &= \frac{1}{2\pi}\int_{-\infty}^{\infty} dt\, G(t)\exp(-i\omega t) \\ &= \langle\langle A; A^+\rangle\rangle_\omega. \end{aligned} \quad (1.66)$$

The equation-of-motion for $G(\omega)$ is then

$$-\omega G(\omega) = \frac{1}{2\pi}\langle[A, A^+]\rangle + \langle\langle[A, \mathcal{H}]; A^+\rangle\rangle_\omega. \quad (1.67)$$

The Green's function on the right-hand side will represent higher-order, collision effects, in all but the simplest cases, and must be approximated in some way.

* Some authors refer to this function as the retarded Green's function.

ISBN 0-8053-6610-5, 0-8053-6611-3 (pbk.)

The relation between $G(\omega)$ and $\tilde{\chi}(i\omega)$ follows immediately from (1.64), namely

$$G(\omega) = -\frac{1}{2\pi}\tilde{\chi}(i\omega). \tag{1.68}$$

The relation between $G(\omega)$ and the scattering function $S(\omega)$ is found by using the following integral representation of the step function,

$$\theta(t) = \frac{i}{2\pi}\int_{-\infty}^{\infty}\frac{du\,\exp(-iut)}{u+i\eta}, \qquad \eta \to 0. \tag{1.69}$$

If we use this result together with (1.37) we find,

$$G(\omega) = \frac{1}{2\pi}\int_{-\infty}^{\infty}\frac{du\,S(u)\{\exp(-u\beta)-1\}}{u-\omega+i\eta} \tag{1.70}$$

and from (1.50) and (1.68) it follows that,

$$G(\omega) = \frac{1}{2\pi^2}\int_{-\infty}^{\infty}\frac{du\,\tilde{\chi}''(iu)}{u-\omega+i\eta} \tag{1.71}$$

$$= -\frac{1}{\pi}\int_{-\infty}^{\infty}\frac{du\,G''(u)}{u-\omega+i\eta}. \tag{1.72}$$

The last result is in the form of a dispersion relation for $G(\omega)$, and it is a direct consequence of the fact that $G(t)$ is, by definition, a causal function. A particularly useful result follows from (1.70) when we use the well-known identity,

$$\lim_{\eta\to 0}\int_{-\infty}^{\infty}\frac{du}{u+i\eta} = -i\pi\delta(u) + P\int_{-\infty}^{\infty}\frac{du}{u} \tag{1.73}$$

where the second term on the right-hand side is the principal part of the integral. If we use (1.73) and exploit the fact that $S(\omega)$ is purely real then it follows from (1.70) that,

$$G''(\omega) = \tfrac{1}{2}\{1-\exp(-\omega\beta)\}S(\omega). \tag{1.74}$$

1.7 EXAMPLE; ANHARMONIC OSCILLATOR

To round off this chapter we consider the calculation of the displacement autocorrelation function for a single atom whose motion is described by the Hamiltonian

$$\mathcal{H} = \tfrac{1}{2}(p^2 + \omega_0^2 u^2) - g_0 u^4. \tag{1.75}$$

Here, the mass of the atom is set equal to unity, p is the momentum conjugate to the displacement u, and g_0 is the strength of the quartic

ISBN 0-8053-6610-5, 0-8053-6611-3 (pbk.)

anharmonic term. The displacement u in (1.75) is assumed to be small, and therefore the quartic term is a small perturbation.

We calculate the displacement autocorrelation function in terms of the Green's function,

$$G(t) = \langle\langle u(t); u\rangle\rangle \tag{1.76}$$

and define its Fourier transform $G(\omega)$ as in (1.66). Using the commutation relation ($m \geq 1$)

$$[u^m, p] = imu^{m-1} \tag{1.77}$$

and the equation-of-motion (1.67) we find

$$-\omega G(\omega) = i\langle\langle p; u\rangle\rangle_\omega \tag{1.78}$$

and, also from (1.67),

$$-i\omega\langle\langle p; u\rangle\rangle_\omega = \frac{1}{2\pi} + \omega_0^2 G(\omega) - 4g_0\langle\langle u^3; u\rangle\rangle_\omega . \tag{1.79}$$

For $g_0 = 0$,

$$\begin{aligned} 2\pi G(\omega) &= (\omega^2 - \omega_0^2)^{-1} \\ &= \frac{1}{2\omega_0}\left\{\frac{1}{\omega - \omega_0} - \frac{1}{\omega + \omega_0}\right\} . \end{aligned} \tag{1.80}$$

The imaginary part of $G(\omega)$ is calculated, according to (1.70), with $\omega \to \omega - i\eta$ and $\eta \to 0$. From (1.80) we find, for $g_0 = 0$,

$$G''(\omega) = \frac{1}{4\omega_0}\{\delta(\omega - \omega_0) - \delta(\omega + \omega_0)\}. \tag{1.81}$$

The corresponding $S(\omega)$ is calculated from (1.74), and we thereby recover the result (1.17) obtained directly from the equation-of-motion for the displacement operator.

If $g_0 \neq 0$ then we are faced with calculating a new Green's function $\langle\langle u^3; u\rangle\rangle_\omega$ in (1.79). There are two almost obvious lines of attack; namely, approximate this new Green's function in terms of $G(\omega)$, or write down an equation-of-motion for the new Green's function and see if a good approximation suggests itself. The latter route is pursued in Chap. V, and for the moment we consider approximating the new Green's function in terms of $G(\omega)$.

If the displacements are weakly correlated then a good approximation is to decouple the new Green's function on the right-hand side of (1.79) in terms of products of $\langle u^2\rangle$ and $G(\omega)$. To make this idea slightly more precise, consider the calculation of $\langle u^4\rangle$ assuming the u's are random variables with a Gaussian distribution as is the case when $g_0 = 0$; then

$$\langle u^4\rangle = 3\langle u^2\rangle^2$$

ISBN 0-8053-6610-5, 0-8053-6611-3 (pbk.)

and, to a good approximation,

$$u^3 \simeq 3\langle u^2 \rangle u.$$

In view of this result we make the approximation

$$\langle\langle u^3; u \rangle\rangle_\omega \simeq 3\langle u^2 \rangle G(\omega). \tag{1.82}$$

The approximation (1.82) is also obtained if we take all possible pairings of u's in the Green's function on the left-hand side. Inserting (1.82) into (1.79), the Green's function $G(\omega)$ is of the same form as (1.80) with the bare frequency replaced by

$$\omega_0 \rightarrow \{\omega_0^2 - 12 g_0 \langle u^2 \rangle\}^{1/2} = \Omega_0. \tag{1.83}$$

A self-consistent equation for $\langle u^2 \rangle$, and hence Ω_0, is obtained from (1.81) using (1.74), namely,

$$\langle u^2 \rangle = \frac{1}{2\Omega_0} \coth(\tfrac{1}{2}\beta\Omega_0). \tag{1.84}$$

In the limit of high temperatures, $\omega_0\beta \rightarrow 0$, and to lowest order in g_0,

$$\langle u^2 \rangle \simeq (\beta\omega_0^2)^{-1}. \tag{1.85}$$

Substituting this approximation in (1.83),

$$\Omega_0 \simeq \omega_0(1 - T/T_c)^{1/2} \tag{1.86}$$

where the critical temperature

$$T_c = \omega_0^4/12 g_0. \tag{1.87}$$

As T approaches T_c, Ω_0 tends to zero (the vibration goes soft) and, in view of (1.84), the mean-square lattice displacement diverges. Our approximate treatment of the quartic anharmonicity in (1.75) leads to the idea that, for $g_0 > 0$, the system is unstable at a high temperature T_c at which the lattice displacement takes on macroscopic values. This idea is central to the soft-mode theory of structural phase transitions [Cochran (1969) and Samara (1977)]. Below T_c an unstable vibration in a crystal is stabilized by higher-order terms in the Hamiltonian of the lattice displacements.

The alert reader will, no doubt, have seen that the result (1.83) can be derived directly from the Hamiltonian, by making the approximation

$$u^4 \simeq 6\langle u^2 \rangle u^2. \tag{1.88}$$

The factor 6 in (1.88) is obtained by counting the number of pairings that are made in the approximation procedure. Once we have made the approximation (1.88) the Hamiltonian is harmonic and the frequency of vibration is given by (1.83), and the result (1.84) follows immediately.

The approximation (1.88) is usually called a mean-field approximation. By making the approximation, we consider the displacement of an atom in

ISBN 0-8053-6610-5, 0-8053-6611-3 (pbk.)

a force field part of which is due to the direct coupling between the atoms, and part of which is due to a temperature dependent, mean-value force that results from the anharmonicity. The approximation scheme is self-consistent, and it leads to a transcendental equation for the mean-square lattice vibration, (1.84). To improve upon this approximation it is necessary to account for fluctuations about the mean value of the force field.

APPENDIX 1A

Let A in (1.28) be of the form

$$A = \theta a + \varphi a^{+} \tag{1A.1}$$

where θ and φ are c-number coefficients and a, a^{+} are Bose operators for a harmonic oscillator of natural frequency ω_0. Using (1.26),

$$\exp A = \exp(\theta a)\cdot\exp(\varphi a^{+})\exp(-\tfrac{1}{2}\theta\varphi)$$

and the calculation of $\langle \exp A \rangle$ therefore reduces to the calculation of

$$\langle \exp(\theta a)\exp(\varphi a^{+})\rangle$$

$$= Z^{-1}\sum_{n=0}^{\infty} \exp(-\beta E_n)\langle n|\exp(\theta a)\exp(\varphi a^{+})|n\rangle, \tag{1A.2}$$

where $E_n = (n + \frac{1}{2})\omega_0$ is the energy eigenvalue for the nth harmonic oscillator state. The partition function,

$$Z = \sum_{n=0}^{\infty} \exp(-\beta E_n) = \exp(\tfrac{1}{2}\omega_0\beta)\, n_0. \tag{1A.3}$$

Using the relation [Ziman (1969)]

$$|n\rangle = (n!)^{-1/2}(a^{+})^{n}|0\rangle$$

we find

$$\exp(\varphi a^{+})|n\rangle = \sum_{m=0}^{\infty}\left\{\frac{(m+n)!}{n!}\right\}^{1/2}\left(\frac{\varphi^{m}}{m!}\right)|m+n\rangle \tag{1A.4}$$

and so

$$\langle n|\exp(\theta a)\cdot\exp(\varphi a^{+})|n\rangle = \sum_{m=0}^{\infty}\frac{(m+n)!}{(m!)^{2}n!}(\theta\varphi)^{m}. \tag{1A.5}$$

In calculating the sum of (1A.5) multiplied by $\exp(-\beta E_n)$, as required in (1A.2), we note that

$$(1-y)^{-1-m} = \sum_{n=0}^{\infty}\frac{(m+n)!}{m!n!}y^{n}.$$

ISBN 0-8053-6610-5, 0-8053-6611-3 (pbk.)

We find,

$$\sum_{n=0}^{\infty} \exp(-\beta E_n)\langle n|\exp(\theta a)\cdot\exp(\varphi a^+)|n\rangle$$

$$= Z\sum_{m=0}^{\infty} [\theta\varphi(1 + n_0)]^m/m!$$

$$= Z\exp\{\theta\varphi(1 + n_0)\}. \tag{1A.6}$$

Inserting this last result in (1A.2) we have, finally,

$$\langle \exp A\rangle = \exp\{\theta\varphi(\langle a^+a\rangle + \tfrac{1}{2})\} \tag{1A.7}$$

which is identical to (1.28).

APPENDIX 1B

From (1.5) and (1.8) we have

$$\langle A^+A(t)\rangle = Z^{-1}\,\mathrm{Tr}\{\exp(-\beta\mathscr{H})A^+\exp(i\mathscr{H}t)A\exp(-i\mathscr{H}t)\}. \tag{1B.1}$$

Using the invariance of the trace in (1B.1) to a cyclic permutation of the operators, the right-hand side of (1B.1) becomes

$$Z^{-1}\,\mathrm{Tr}\{\exp(i\mathscr{H}t)A\exp[-i\mathscr{H}(t - i\beta)]A^+\}$$

$$= Z^{-1}\,\mathrm{Tr}\{\exp(-\beta\mathscr{H})\exp[i\mathscr{H}(t - i\beta)]A\exp[-i\mathscr{H}(t - i\beta)]A^+\}$$

and this is equivalent to the right-hand side of (1.37).

APPENDIX 1C

From the definitions (1.43) and (1.51a) we find that the relaxation function is

$$R(t) = i\int_t^{\infty} d\bar{t}\,\langle A(\bar{t})A^+ - A^+A(\bar{t})\rangle. \tag{1C.1}$$

It is convenient in the following development to replace the upper limit in the integral by a time T that will subsequently be allowed to go to infinity. Using (1.37), the right-hand side of (1C.1) becomes

$$i\left\{\int_{t+i\beta}^{T+i\beta} - \int_t^T\right\} d\bar{t}\,\langle A^+A(\bar{t})\rangle. \tag{1C.2}$$

The form of this result suggests that we consider the integral of the correlation function $\langle A^+A(\bar{t})\rangle$ along the path shown in Fig. (1.1) in the upper-half of the complex plane. We can easily prove that $\langle A^+A(\bar{t})\rangle$ is well

ISBN 0-8053-6610-5, 0-8053-6611-3 (pbk.)

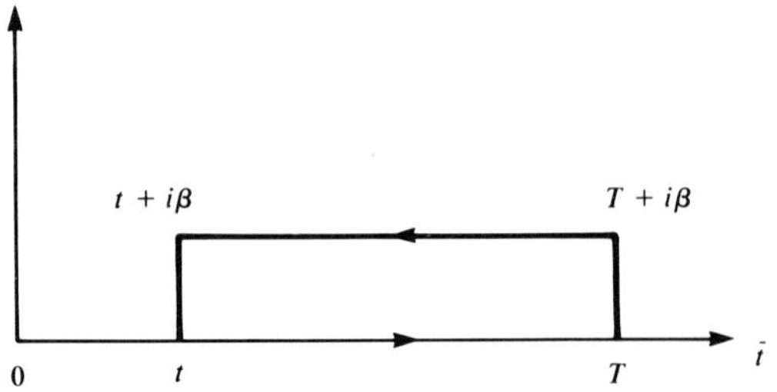

Fig. 1.1 The diagram depicts the contour in the upper-half plane used in Appendix C.

behaved in this region, since

$$\langle A^+A(t+i\epsilon)\rangle = Z^{-1}\sum_{\lambda\lambda'} |A_{\lambda'\lambda}|^2 \exp\{-(\beta-\epsilon)E_\lambda + i\bar{t}(E_{\lambda'}-E_\lambda) - \epsilon E_{\lambda'}\}$$

and provided that $\epsilon \leq \beta$ the sums converge. The integral of $\langle A^+A(\bar{t}+i\epsilon)\rangle$ round the contour in Fig. (1.1) is therefore zero. Hence (1C.2) reduces to

$$i\left\{\int_T^{T+i\beta} + \int_{t+i\beta}^{t}\right\} d\bar{t}\,\langle A^+A(\bar{t})\rangle$$

$$= \int_0^\beta d\tau\,\langle A^+A(t+i\tau)\rangle - \int_0^\beta d\tau\,\langle A^+A(T+i\tau)\rangle$$

and

$$R(t) = \int_0^\beta d\tau\,\langle A^+A(t+i\tau)\rangle - \lim_{T\to\infty}\int_0^\beta d\tau\,\langle A^+A(T+i\tau)\rangle$$

$$= \int_0^\beta d\tau\,\langle\{A^+(-i\tau) - \langle A^+\rangle\}\{A(t) - \langle A\rangle\}\rangle \qquad (1C.3)$$

where in obtaining the second equality we have appealed to the law of increase in entropy and used (1.54).

ISBN 0-8053-6610-5, 0-8053-6611-3 (pbk.)

CHAPTER II

STOCHASTIC EQUATIONS

A common feature of this chapter and Chaps. III and IV is an emphasis on properties at large distances and long times, and often referred to as the macroscopic properties of a system. In Sec. 2.1 we consider a simple model of stochastic (Brownian) motion, and the calculation of the self-diffusion coefficient is used to introduce the discussion of transport coefficients given in Sec. 2.2. A nonlinear model of an incompressible liquid is analyzed in detail in Sec. 2.3. Having discussed two simple models we review, in Sec. 2.4, the salient features of stochastic models and discuss the compatibility of their static and dynamic properties. The nonlinearity in the model incompressible liquid discussed in Sec. 2.3 is quadratic in the velocity variables. Models with cubic nonlinearities are often of interest and we therefore conclude the chapter with a brief discussion of an example of such a model derived from the general considerations given in Sec. 2.4.

All the model systems considered in this chapter are obtained by considering the macroscopic motion of the variables of interest (continuum mechanics), and motions on an atomic (microscopic) scale are included in the guise of a random, stirring or agitating force [Fox (1978)]. Because the characteristic time scales associated with the dynamical variables and random forces are very different their motions are essentially uncorrelated, and random forces do not therefore appear explicitly in the autocorrelation functions of macroscopic variables. The random force introduced in stochastic equations is meant to model that part of the dynamics which is neglected in continuum equations-of-motion, and since it is largely arbitrary in form it should not appear explicitly in a measurable quantity.

Attempts to go beyond stochastic equations, and construct equations-of-motions that are free from the restriction of large distances and long times, have been made successfully in terms of a generalized Langevin equation discussed in Chap. III. While it is always possible to obtain simple

Stephen Lovesey, Condensed Matter Physics: Dynamic Correlations

ISBN 0-8053-6610-5, 0-8053-6611-3 (pbk.)

stochastic equations from the more general and sophisticated theory of Chap. III it is nonetheless worthwhile to consider separately some examples of these equations and their properties. A well-known example of this type is based on linear hydrodynamic equations-of-motion for density fluctuations in a compressible liquid [Landau and Lifshitz (1959) and (1963)], and since this model has been treated by several authors we shall not discuss it here but we use it as an example in Chap. III of how to obtain correct hydrodynamic equations from a generalized Langevin equation.

The basic differences between the dynamic properties of gases, solids and liquids may be attributed to the differences in the characteristic time, and energy, scales for the three states. For a dilute gas the important time scales are, in order of increasing magnitude, the collision time τ_c and the mean-free time τ_m, and the ratio of the average potential and kinetic energies is much less than unity. Processes that occur on a time scale $\ll \tau_c$ must be described by microscopic equations, whereas processes that occur on a time scale $\gg \tau_m$ are appropriately described by continuum or hydrodynamic equations. The intermediate time region is usually referred to as the kinetic regime. In a solid the magnitude of the two characteristic time scales is reversed compared to a dilute gas, and the average potential energy exceeds the average kinetic energy. The difficulties encountered in describing the dynamic properties of a dense liquid can be ascribed to the fact that there is no distinct separation between the time and energy scales [Hansen and McDonald (1976), and Résibois and de Leener (1977)].

Simple phenomenological theories of dynamic effects in liquids lead to an exponential decay of correlated motion with increasing time. We shall find, however, that a sophisticated approximation for nonlinear effects predicts in addition an inverse power law decrease of correlated motion. The existence of power law tails in correlation functions narrows the difference between macroscopic and microscopic characteristic times and precludes the application of hydrodynamic equations to phenomena that vary on a long enough time scale. Obviously the difficulties increase near the critical point where the growth of characteristic distances is accompanied by a decrease of characteristic times [Pomeau and Résibois (1975)].

2.1 LANGEVIN EQUATION

The (linear) Langevin equation for the motion of an atom in a liquid is a simple example of a stochastic equation-of-motion, and we consider some of its properties to set the stage for the study of nonlinear stochastic equations.

A given atom in a liquid undergoes many collisions with the other atoms in the liquid due to thermal agitation. When viewed over a time interval much greater than that between collisions it appears to move slowly, its velocity v being determined by the equation of (Brownian) motion [Chan-

ISBN 0-8053-6610-5, 0-8053-6611-3 (pbk.)

drasekhar (1943), MacDonald (1962), Sargent *et al.* (1974), and Fox (1978)]

$$\partial_t v(t) = -\gamma v(t) + f(t). \tag{2.1}$$

The frictional force exerted by the liquid is represented by the first term on the right-hand side of (2.1), and the second term, $f(t)$, represents the random force due to the random collisions with the other atoms. Because γ and $f(t)$ have the same physical origin, namely the motion of the other atoms in the liquid, they are intimately related to each other, as we shall shortly verify.

By definition the average value of the random force is zero. It is assumed, moreover, that $f(t)$ is uncorrelated with the velocity at earlier times, so that $\langle f(t)v(t_0)\rangle = 0$ for $t \geq t_0$, and also to have a white spectrum

$$\langle f(t)\, f(t_0)\rangle \propto \delta(t - t_0). \tag{2.2}$$

Here, the averaging of the products of variables is performed over a time interval which is long compared to the rapid variations in $f(t)$ and short compared to the damping time γ^{-1}. The relation (2.2) simply means that the values of $f(t)$ at different times are completely uncorrelated. The constant of proportionality in (2.2) can be determined by requiring that $\langle v(t)^2\rangle = \langle v(0)^2\rangle$ because the liquid system is stationary. From the equation-of-motion (2.1),

$$v(t) = \exp(-\gamma t)\left\{ v + \int_0^t d\bar{t}\, f(\bar{t}) \exp(\gamma \bar{t})\right\} \tag{2.3}$$

and so

$$\begin{aligned}\langle v(t)^2\rangle &= \exp(-2\gamma t)\\ &\quad\times \left\{\langle v^2\rangle + \int_0^t d\bar{t} \int_0^t d\bar{t}'\, \langle f(\bar{t})\, f(\bar{t}')\rangle \exp\{\gamma(\bar{t} + \bar{t}')\}\right\}\\ &= \langle v^2\rangle \end{aligned} \tag{2.4}$$

which gives

$$\langle f(t)\, f(t_0)\rangle = 2\gamma\langle v^2\rangle\delta(t - t_0) = (2\gamma/M\beta)\delta(t - t_0), \tag{2.5}$$

where M is the mass of the particle. The second equality in (2.5) follows if we assume the law of equipartition of energy. Equation (2.5) is the desired relation between the frictional coefficient γ and the random force, $f(t)$.

To obtain the velocity autocorrelation function we multiply (2.3) by $v(0)$, and average the resulting equation, noting that $v(0)$ and $f(t)$ are uncorrelated,

$$\langle vv(t)\rangle = \langle v^2\rangle \exp(-\gamma|t|). \tag{2.6}$$

This result cannot hold for small t because the autocorrelation function in

ISBN 0-8053-6610-5, 0-8053-6611-3 (pbk.)

reality cannot have a discontinuous gradient at $t = 0$ as (2.6) implies. In fact, it can be shown that for small t [Nijboer and Rahman (1966)]

$$\langle vv(t)\rangle = \langle v^2\rangle\{1 - \Omega_0^2 t^2/2 + \cdots\} \tag{2.7}$$

where the frequency Ω_0 is often referred to as the Einstein frequency of vibration. One of the main objectives of the theory described in Chap. III is to construct stochastic equations-of-motion which properly reproduce the short-time behavior of autocorrelation functions.

Before leaving the simple Langevin equation, we derive an expression for the self-diffusion constant D as the integral of the velocity autocorrelation function. A simple Langevin equation for the conductivity of a superionic material is discussed by Boyce and Huberman (1979).

The self-diffusion constant is usually defined by the relation

$$D = \frac{1}{3}\lim_{t\to\infty}\frac{1}{2t}\langle\{\mathbf{R}(0) - \mathbf{R}(t)\}^2\rangle \tag{2.8}$$

where $\mathbf{R}(t)$ is the position of the atom at time t. The mean-square displacement $\langle R^2(t)\rangle$ of atoms in a computer simulation experiment is shown in Fig. (2.1). In the model, simulated pairs of atoms interact through a Lennard-Jones potential, and the temperature and density are chosen so that the model is representative of liquid argon near its triple point. A linear time dependence is achieved at $t \sim 1.5 \times 10^{-12}$ sec, and the diffusion constant deduced from the limiting slope ($= 6D/\sigma^2$, where σ is the position at which the potential is zero) is 1.8×10^{-5} cm^2 sec^{-1}.

In order to express D as an integral of the velocity autocorrelation

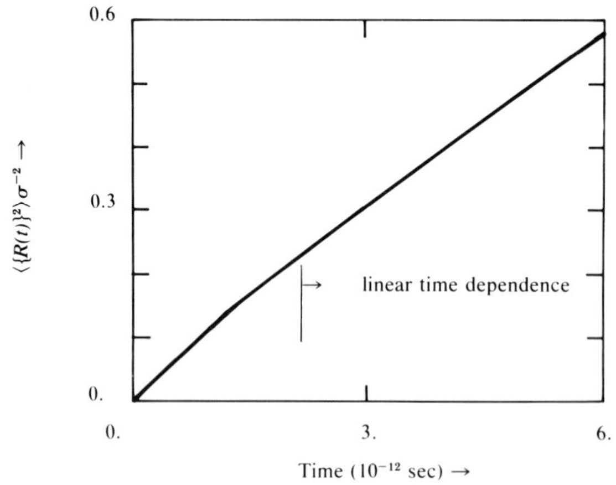

Fig. (2.1) The mean-square displacement $\langle R(t)^2\rangle$ for a Lennard-Jones liquid is shown in units of σ^2, where σ is the position of the zero in the potential. [After Levesque and Verlet (1970).]

ISBN 0-8053-6610-5, 0-8053-6611-3 (pbk.)

function we write

$$\mathbf{R}(t) - \mathbf{R}(0) = \int_0^t d\bar{t}\, \mathbf{v}(\bar{t}),$$

and then,

$$\begin{aligned}\langle\{\mathbf{R}(t) - \mathbf{R}(0)\}^2\rangle &= \left\langle\left\{\int_0^t d\bar{t}\, \mathbf{v}(\bar{t})\right\}^2\right\rangle \\ &= 2\int_0^t d\bar{t}\int_0^{\bar{t}} d\bar{t}'\, \langle \mathbf{v}(\bar{t})\cdot\mathbf{v}(\bar{t}')\rangle \\ &= 2\int_0^t d\bar{t}\, (t - \bar{t})\langle \mathbf{v}\cdot\mathbf{v}(\bar{t})\rangle .\end{aligned}$$

From this last result, and the definition (2.8) we obtain the desired result,

$$D = \frac{1}{3}\int_0^\infty dt\, \langle \mathbf{v}\cdot\mathbf{v}(t)\rangle . \tag{2.9}$$

If the velocity autocorrelation function is assumed to be given by (2.6) for all times, then

$$D = \langle v^2\rangle/\gamma = (M\beta\gamma)^{-1}, \tag{2.10}$$

which is the familiar relation between the diffusion constant and frictional coefficient derived by Einstein (1910).

2.2 TRANSPORT COEFFICIENTS

The result (2.9) is a simple example of more general results derived by Green (1952) and Kubo [(1957) and (1966)] that relate transport coefficients to integrals of appropriate autocorrelation functions. Transport coefficients govern the time evolution of processes that occur on macroscopic length and time scales, e.g. self-diffusion of a Brownian particle, and electric conduction. The variable associated with a transport coefficient has the special property of being a constant-of-motion in the long-wavelength limit, and such variables are usually called conserved variables. A conserved variable is, therefore, associated with a macroscopic process in the long-wavelength limit. In consequence, its equation-of-motion takes the form of a conservation, or continuity, equation in which the time derivative of the conserved variable is proportional to the divergence of a flux, or current density. The spatial Fourier components of a conserved variables φ_{k} satisfy an equation

$$\partial_t \varphi_{\mathrm{k}} + i\,\mathbf{k}\cdot\mathbf{J}_{\mathrm{k}} = 0 \tag{2.11}$$

ISBN 0-8053-6610-5, 0-8053-6611-3 (pbk.)

where $\mathbf{J}_\mathbf{k}$ is the Fourier transform of the flux, and, since $\varphi_\mathbf{k}$ is a conserved variable,

$$\lim_{\mathbf{k}\to 0} \partial_t \varphi_\mathbf{k} = 0. \tag{2.12}$$

When the actual form of the flux in (2.11) is derived from macroscopic considerations the resulting equation is valid for wavevectors **k** that do not exceed an upper limit which is of the order of the inverse of a molecular length.

The macroscopic properties of the system described by $\varphi_\mathbf{k}$ lead us to anticipate that, in the limit of small **k** and long times, the autocorrelation function of $\varphi_\mathbf{k}$ takes the form

$$\langle \varphi_\mathbf{k}^* \varphi_\mathbf{k}(t) \rangle \sim \exp(-k^2 L t) \quad \text{macroscopic limit} \tag{2.13}$$

where L denotes the appropriate transport coefficient. A relation between L and the autocorrelation function of the flux associated with $\varphi_\mathbf{k}$, analogous to the relation (2.9) for self-diffusion, is obtained by the following argument. From (2.11) and (1.39),

$$-\partial_t^2 \langle \varphi_\mathbf{k}^* \varphi_\mathbf{k}(t) \rangle = \langle \mathbf{k}\cdot\mathbf{J}_\mathbf{k}^* \mathbf{k}\cdot\mathbf{J}_\mathbf{k}(t) \rangle$$

and for a spatially isotropic system the right-hand side is equal to, $\frac{1}{3}k^2 \langle \mathbf{J}_\mathbf{k}^* \cdot \mathbf{J}_\mathbf{k}(t) \rangle$. We then find

$$\langle \varphi_\mathbf{k}^* \varphi_\mathbf{k}(t) \rangle = \langle |\varphi_\mathbf{k}|^2 \rangle - \tfrac{1}{3}k^2 \int_0^t d\bar{t}\, (t - \bar{t}) \langle \mathbf{J}_\mathbf{k}^* \cdot \mathbf{J}_\mathbf{k}(\bar{t}) \rangle$$

and so, if the transport coefficient exists, it is given by

$$L = \frac{1}{3} \int_0^\infty dt \lim_{\mathbf{k}\to 0} \{ \langle \mathbf{J}_\mathbf{k}^* \cdot \mathbf{J}_\mathbf{k}(t) \rangle / \langle |\varphi_\mathbf{k}|^2 \rangle \}. \tag{2.14}$$

We stress that in (2.14) the limit $\mathbf{k} \to 0$ must be taken before the integral is evaluated. For, since we assume a complete loss of correlation between the variables $\varphi_\mathbf{k}^*(0)$ and $\varphi_\mathbf{k}(t)$ as $t \to \infty$, so that $\lim_{t\to\infty} \langle \varphi_\mathbf{k}^* \varphi_\mathbf{k}(t) \rangle = 0$, then for finite **k**, $\int_0^\infty dt\, \langle \mathbf{J}_\mathbf{k}^* \cdot \mathbf{J}_\mathbf{k}(t) \rangle = 0$. Notice also that, with a suitable choice of wavevector independent variables, the result (2.14) for the transport coefficient can be cast into the same form as (2.8).

Let us now consider the case of self-diffusion, and demonstrate that the foregoing argument reproduces the result (2.9). For a single particle, whose position is defined by the vector $\mathbf{R}(t)$, the particle density function in real space is $\delta(\mathbf{r} - \mathbf{R})$, and its spatial Fourier transform

$$\begin{aligned} \varphi_\mathbf{k} &= \int d\mathbf{r}\, \delta(\mathbf{r} - \mathbf{R}(t)) \exp(i\mathbf{k}\cdot\mathbf{r}) \\ &= \exp(i\mathbf{k}\cdot\mathbf{R}(t)). \end{aligned} \tag{2.15}$$

ISBN 0-8053-6610-5, 0-8053-6611-3 (pbk.)

Hence, $i\partial_t\varphi_{\mathbf{k}} = -\mathbf{k}\cdot\mathbf{v}\varphi_{\mathbf{k}}$ where $\mathbf{v}$ is the velocity of the particle. From this result it follows that, for single particle motion

$$\mathbf{J}_{\mathbf{k}} = -\mathbf{v}\varphi_{\mathbf{k}}. \tag{2.16}$$

Using (2.16) in (2.14) we recover the result (2.9) with the transport coefficient L in (2.14) identified as the self-diffusion coefficient, D.

It is worthwhile, perhaps, to comment further on the relation between the continuity equation and the underlying macroscopic process. In thermal equilibrium the average value of a conserved variable is independent of the spatial and time coordinates, and the average value of the corresponding flux is zero. If, now, the system is disturbed slightly, so that it deviates a little from the true equilibrium state, we might expect the fluctuation in the conserved variable away from its equilibrium value to satisfy a continuity equation in which the nonequilibrium average value of the flux is proportional to the gradient of the conserved variable. For example, in the case of self-diffusion in a liquid, the nonequilibrium average value of the particle current density $= -D\nabla\bar{\varphi}(\mathbf{r}, t)$ where $\bar{\varphi}(\mathbf{r}, t)$ is the nonequilibrium average value of the particle density, and the minus sign indicates that the net flow is in the direction in which $\bar{\varphi}$ decreases since D is assumed positive. The continuity equation for $\bar{\varphi}$ is now $(\partial_t - D\nabla^2)\bar{\varphi}(\mathbf{r}, t) = 0$ which is the familiar Fick's law for self-diffusion. If we introduce the Laplace transform of $\bar{\varphi}_{\mathbf{k}}(t)$, then Fick's law leads to the result

$$\int_0^\infty dt\; \bar{\varphi}_{\mathbf{k}}(t)\exp(-st) = \bar{\varphi}_{\mathbf{k}}(t=0)\{s + Dk^2\}^{-1},$$

and the diffusion process manifests itself here as a pole at the frequency $\omega = iDk^2$ which vanishes in the long-wavelength limit.

The connection between what we have just said and the earlier discussion in this section is the assumption that the long-wavelength, *spontaneous* equilibrium fluctuations in the self-motion of a particle are governed by an equation of the same form of Fick's law, and that, in particular, the particle density autocorrelation function satisfies the equation,

$$(\partial_t + k^2D)\langle\varphi_{\mathbf{k}}^*\varphi_{\mathbf{k}}(t)\rangle = 0.$$

The physical solution of this latter equation has the same exponential time dependence as was assumed in Eq. (2.13) for the macroscopic limit of the autocorrelation function.

2.3 INCOMPRESSIBLE NAVIER-STOKES LIQUID

Equation (2.1) for the velocity $\mathbf{v}$, and the definition of the random force f lead to an exactly solvable problem. A more realistic equation-of-motion contains nonlinear terms in the velocity and, in general, it will not admit an

ISBN 0-8053-6610-5, 0-8053-6611-3 (pbk.)

exact solution. We consider such a case in this section, where we examine the long-wavelength properties of an incompressible liquid.

The equation-of-motion in a continuum of mass density ρ takes the simple form [Landau and Liftshitz (1963)]

$$\rho(\partial_t + \mathbf{v}\cdot\boldsymbol{\nabla})\mathbf{v} = \boldsymbol{\nabla}\cdot\boldsymbol{\sigma}, \quad \text{with } (\boldsymbol{\nabla}\cdot\boldsymbol{\sigma})_\alpha = \sum_\beta \partial_{x_\beta}\sigma_{\beta\alpha}.$$

Here α and β denote Cartesian vector components, and σ is the stress tensor. The operator acting on $\mathbf{v}$ on the left-hand side is the time-change as seen by an observer who moves with the liquid, i.e. whose velocity is always $\mathbf{v}$ whatever position he may have been brought to. The short-time and short-distance forces operating in a realistic model of a liquid and neglected in the continuum equation-of-motion are modelled by a random force f which is added to the stress force. Hence, our model equation-of-motion is,

$$\rho(\partial_t + \mathbf{v}\cdot\boldsymbol{\nabla})\mathbf{v} = \boldsymbol{\nabla}\cdot\boldsymbol{\sigma} + \mathbf{f}. \tag{2.17}$$

The properties of most liquids are well described by a stress tensor with a single parameter η called the coefficient of viscosity, and

$$\boldsymbol{\sigma} = -p\mathbf{I} + 2\eta\boldsymbol{\sigma}' \tag{2.18}$$

where $\mathbf{I}$ is the unit tensor, the pressure

$$p = -\tfrac{1}{3}\operatorname{Tr}\sigma \tag{2.19}$$

and

$$\sigma'_{\alpha\beta} = \tfrac{1}{2}(\partial_{x_\beta}v_\alpha + \partial_{x_\alpha}v_\beta) - \tfrac{1}{3}\delta_{\alpha,\beta}(\boldsymbol{\nabla}\cdot\mathbf{v}). \tag{2.20}$$

Using (2.18) in (2.17) leads to the Navier-Stokes equation-of-motion

$$\rho(\partial_t + \mathbf{v}\cdot\boldsymbol{\nabla})\mathbf{v} = \mathbf{f} + \eta\{\nabla^2\mathbf{v} + \tfrac{1}{3}\boldsymbol{\nabla}(\boldsymbol{\nabla}\cdot\mathbf{v})\} - \boldsymbol{\nabla}p. \tag{2.21}$$

For an incompressible liquid,* the conservation equation for the particle density reduces to,

$$\boldsymbol{\nabla}\cdot\mathbf{v} = 0, \tag{2.22}$$

and in this case the nonlinear term in (2.21) arises from purely kinematic considerations, or put another way it arises from Galilean invariance. The dynamics of the model are specified by (2.21) and (2.22), and $\mathbf{f}$ is taken to be a Gaussian-distributed random force uncorrelated with $\mathbf{v}$, just as in Sec. 2.1. The random force acts as a source of energy to balance the dissipative effect of viscosity, and it thereby makes a stationary state possible. It should be stressed that $\mathbf{f}$ is not a real force; it is a hypothetical, isotropic and homogeneous stirring force that mimics the real external forces which

* The velocity of sound in an incompressible liquid is infinite.

ISBN 0-8053-6610-5, 0-8053-6611-3 (pbk.)

enable a stationary state to be established in a liquid. Because (2.21) is based upon a continuum model the range of lengths involved are restricted to be larger than a typical molecular separation.

Equilibrium values of the velocity components are assumed to be given by an equilibrium distribution function of the form

$$P_0 = \exp\left(-\frac{1}{2\chi}\int d\mathbf{r}\ \mathbf{v}^2(\mathbf{r})\right). \tag{2.23}$$

If the distribution is Boltzmann then the susceptibility χ is proportional to temperature and inversely proportional to the mass density, and it has the dimension of $(\text{length})^{d+2}/(\text{time})^2$ where d = number of space dimensions. From (2.23) it follows that the average velocity is zero. This must be true for an isotropic liquid, since a nonzero average velocity would specify a unique spatial direction.

It is convenient to write the velocity in terms of dimensionless Fourier components $\mathbf{a}_\mathbf{k}$, where the wavevectors $\mathbf{k}$ have a density $\Omega/(2\pi)^d$ and Ω = volume of the liquid. The maximum value of $\mathbf{k}$ is less than the inverse of a typical molecular length. With the definition,

$$\mathbf{v}(\mathbf{r}) = \frac{(\Omega\chi)^{1/2}}{(2\pi)^d}\int d\mathbf{k}\ \mathbf{a}_\mathbf{k}\exp(-i\mathbf{k}\cdot\mathbf{r}) \tag{2.24}$$

we have $\mathbf{a}_\mathbf{k} = (\mathbf{a}_{-\mathbf{k}})^*$ since $\mathbf{v}(\mathbf{r})$ is real, and

$$\int d\mathbf{r}\ v_\alpha(\mathbf{r})v_\beta(\mathbf{r}) = \frac{\Omega\chi}{(2\pi)^{2d}}\int d\mathbf{r}\int d\mathbf{k}\int d\mathbf{q}\ a_\mathbf{k}^\alpha a_\mathbf{q}^\beta \exp\{-i\mathbf{r}\cdot(\mathbf{k}+\mathbf{q})\}$$

$$= \frac{\Omega\chi}{(2\pi)^d}\int d\mathbf{k}\int d\mathbf{q}\ a_\mathbf{k}^\alpha\, a_\mathbf{q}^\beta\, \delta(\mathbf{k}+\mathbf{q})$$

$$= \chi\sum_\mathbf{k} a_\mathbf{k}^\alpha a_{-\mathbf{k}}^\beta$$

so that the distribution function takes the simple form,

$$P_0 = \exp\{-\tfrac{1}{2}\sum_\mathbf{k}|\mathbf{a}_\mathbf{k}|^2\}. \tag{2.25}$$

The incompressibility condition (2.22) leads to the relation,

$$\sum_\alpha k_\alpha a_\mathbf{k}^\alpha = 0. \tag{2.26}$$

The equation-of-motion satisfied by the Fourier components $\mathbf{a}_\mathbf{k}$ is obtained from (2.21) and (2.22). It is possible, by virtue of the incompressibility condition, to eliminate the pressure from the equation-of-motion. There are several different methods by which this can be done. Here we proceed by noting that, for $\mathbf{f} = 0$, Poisson's equation, obtained from (2.21) and (2.22),

$$\nabla^2 p = -\rho\sum_{\alpha\beta}\partial_{x_\alpha}v_\beta\partial_{x_\beta}v_\alpha$$

ISBN 0-8053-6610-5, 0-8053-6611-3 (pbk.)

leads to

$$p_{\mathrm{k}} = \frac{1}{(2\pi)^d} \int d\mathbf{r}\; p(\mathbf{r}) \exp(i\mathbf{k}\cdot\mathbf{r})$$

$$= -\frac{\rho\chi}{k^2(2\pi)^d} \sum_{\mathrm{q}} \sum_{\alpha\beta} k_\alpha q_\beta a^\alpha_{\mathrm{q}} a^\beta_{\mathrm{k-q}}.$$

From this result it is seen that the pressure contributes a quadratic interaction to the equation-of-motion for $\mathbf{a}_{\mathrm{k}}$, which is conveniently combined with the quadratic interaction arising from Galilean invariance. Assembling the various terms we find that the equation-of-motion for a^α_{k} can be written in the compact form,

$$(\partial_t + \nu_0 k^2) a^\alpha_{\mathrm{k}} = i(\chi/\Omega)^{1/2} \sum_{\beta\gamma} \sum_{\mathrm{pq}} M^{\alpha\beta\gamma}_{\mathrm{k}}(\mathbf{p}, \mathbf{q}) a^\beta_{\mathrm{p}} a^\gamma_{\mathrm{q}} + f^\alpha_{\mathrm{k}}. \tag{2.27}$$

Here,

$$M^{\alpha\beta\gamma}_{\mathrm{k}}(\mathbf{p}, \mathbf{q}) = k_\beta \delta_{\mathrm{k,p+q}} D^{\alpha\gamma}_{\mathrm{k}} \tag{2.28}$$

where

$$D^{\alpha\gamma}_{\mathrm{k}} = \delta_{\alpha\gamma} - \hat{k}_\alpha \hat{k}_\gamma, \quad \text{with } \hat{\mathbf{k}} = \mathbf{k}/k. \tag{2.29}$$

A definition of the random force f^α_{k} which is compatible with our original definition gives

$$\langle f^\alpha_{\mathrm{p}} f^\beta_{-\mathrm{q}}(t) \rangle = 2p^2 \nu_0 \delta_{\mathrm{p,q}} D^{\alpha\beta}_{\mathrm{p}} \delta(t). \tag{2.30}$$

The wavevector squared occurs in (2.30) because the velocity is a conserved variable. The quantity $\nu_0 = (\eta/\rho)$ is usually called the kinematic viscosity.

Equation (2.27) is a nonlinear stochastic equation for the dimensionless Fourier components of the velocity field. The quadratic interaction term serves to put energy into, or to take energy out of, the velocity mode with wavevector **k**. The total wavevector is conserved in the interaction by virtue of the Kronecker delta function in $M^{\alpha\beta\gamma}_{\mathrm{k}}(\mathbf{p}, \mathbf{q})$, Eq. (2.28). The investigation of the effect of the quadratic interaction on the long-wavelength properties of the autocorrelation function $\langle \mathbf{a}^*_{\mathrm{k}}\cdot\mathbf{a}_{\mathrm{k}}(t)\rangle$ is the central task of this section. Notice that for the motions in the liquid to be homogeneous, i.e. uniform in space, the total wavevector in the autocorrelation function must add up to zero.

If we integrate (2.27) then,

$$a^\alpha_{\mathrm{k}}(t) = \exp(-\nu_0 k^2 |t|) \left\{ a^\alpha_{\mathrm{k}} + \int_0^t d\bar{t} \exp(\nu_0 k^2 |\bar{t}|) \left[i \left(\frac{\chi}{\Omega}\right)^{1/2} \right.\right.$$

$$\left.\left. \times \sum_{\beta\gamma} \sum_{\mathrm{pq}} M^{\alpha\beta\gamma}_{\mathrm{k}}(\mathbf{p}, \mathbf{q}) a^\beta_{\mathrm{p}}(\bar{t}) a^\gamma_{\mathrm{q}}(\bar{t}) + f^\alpha_{\mathrm{k}}(\bar{t}) \right] \right\}. \tag{2.31}$$

An equation-of-motion for the autocorrelation function is obtained from

ISBN 0-8053-6610-5, 0-8053-6611-3 (pbk.)

(2.27) by multiplying from the right by $a^{\alpha'}_{-\mathrm{k}}$ and averaging both sides of the resulting equation,

$$(\partial_t + \nu_0 k^2)\langle a^{\alpha}_{\mathrm{k}}(t) a^{\alpha'}_{-\mathrm{k}}\rangle = i\left(\frac{\chi}{\Omega}\right)^{1/2} \sum_{\beta\gamma} \sum_{\mathrm{pq}} M^{\alpha\beta\gamma}_{\mathrm{k}}(\mathbf{p}, \mathbf{q})\langle a^{\beta}_{\mathrm{p}} a^{\gamma}_{\mathrm{q}} a^{\alpha'}_{-\mathrm{k}}(-t)\rangle \tag{2.32}$$

where we have used the fact that the components $\mathbf{a}_{\mathrm{k}}$ and the random force are uncorrelated.

Because the equilibrium distribution function is an even function of the Fourier components it follows that the average value of an odd number of components with the same value of t must vanish. The right-hand side of (2.32) is therefore zero at $t = 0$.

If we form the equation-of-motion for $a^{\alpha'}_{-\mathrm{k}}(-t)$ from (2.31) and substitute the result in (2.32) then we find

$$\begin{aligned}(\partial_t + \nu_0 k^2)&\langle a^{\alpha}_{\mathrm{k}}(t) a^{\alpha'}_{-\mathrm{k}}\rangle \\ &= -\frac{\chi}{\Omega} \sum_{\beta\gamma} \sum_{\mathrm{pq}} M^{\alpha\beta\gamma}_{\mathrm{k}}(\mathbf{p}, \mathbf{q}) \sum_{\beta'\gamma'} \sum_{\mathrm{p'q'}} M^{\alpha'\beta'\gamma'}_{\mathrm{k}}(\mathbf{p}', \mathbf{q}') \\ &\quad \times \int_0^t d\bar{t}\, \langle a^{\beta}_{\mathrm{p}}(\bar{t}) a^{\gamma}_{\mathrm{q}}(\bar{t}) a^{\beta'*}_{\mathrm{p'}} a^{\gamma'*}_{\mathrm{q'}}\rangle \exp(-\nu_0 k^2|t| + \nu_0 k^2|\bar{t}|).\end{aligned} \tag{2.33}$$

This equation is exact, and it forms a convenient starting point to form useful approximate results for the autocorrelation function.

An almost "obvious" approximation to consider follows the reasoning given in Sec. 1.7 in the discussion of anharmonic contributions to the oscillatory properties of a particle. Applied to the present problem, this approximation amounts to decomposing the fourth-order correlation function in (2.33) into products of autocorrelation functions. Of course, this means that we neglect some part of the correlated motion contained in the fourth-order correlation function. We note that the exponential that multiplies the fourth-order correlation function in the integrand is, in fact, a noninteracting, or bare, correlation function, with a time argument $t - \bar{t}$. Hence, we might hope to offset the error, introduced by decomposing the fourth-order correlation function, to some extent by replacing the bare correlation function by a full, interacting correlation function with a time argument $t - \bar{t}$. Making these two steps, namely decomposing the correlation function and replacing the bare correlation function with a full correlation function, results in a self-consistent equation for the desired autocorrelation function, and, in this respect, our proposed approximation scheme is like the mean-field approximation discussed in Sec. 1.7.

The approximation scheme just proposed is not quite so easy to implement as might have been implied. For one thing, we have not given a detailed prescription for the decomposition of the fourth-order correlation function, which amounts essentially to approximating the correlation func-

ISBN 0-8053-6610-5, 0-8053-6611-3 (pbk.)

tion by products of pairs, or autocorrelation functions. Of the various pairs of variables that can be selected we exclude pairs with the same time argument, since such terms are independent of time for a stationary system.

The decomposition proposed would be perfectly correct if the dynamical variables obeyed a Gaussian distribution as is the case for a particle in a harmonic potential. In view of this we can formulate the mechanics of the proposed decomposition by considering the calculation of a multiple correlation function of Bose operators for a harmonic oscillator; the Bose annihilation operator a can be associated with $a_{\mathrm{k}}(t)$, and the creation operator a^+ associated with $a_{\mathrm{p}}(0)$, and since $\langle a^+a^+\cdots a^+\rangle = \langle aa\cdots a\rangle = 0$ we exclude terms with the same time argument. Consider then the calculation of $\langle a^l a^{+l}\rangle$ where l is a positive integer. It is easy to prove, using explicit forms for the harmonic oscillator states, that

$$\langle a^l a^{+l}\rangle = l!\langle aa^+\rangle^l.$$

For the present discussion, it is important to note that the same result is obtained by taking all possible pairings of the operators on the left-hand side and dividing the result by $l!$. Consider the case of immediate interest for which $l = 2$, then the pairing prescription gives,

$$\langle a^2 a^{+2}\rangle = \frac{1}{2!}\langle aaa^+a^+\rangle = 2\langle aa^+\rangle^2,$$

in agreement with the result obtained from the previous, general result setting $l = 2$. With these results in mind we approximate the fourth-order correlation function in (2.33) as follows;

$$\begin{aligned}\langle a_{\mathrm{p}}^{\beta}(t)a_{\mathrm{q}}^{\gamma}(t)a_{-\mathrm{p}'}^{\beta'}a_{-\mathrm{q}'}^{\gamma'}\rangle &\simeq \langle a_{\mathrm{p}}^{\beta}(t)a_{-\mathrm{p}'}^{\beta'}\rangle\langle a_{\mathrm{q}}^{\gamma}(t)a_{-\mathrm{q}'}^{\gamma'}\rangle \\ &\quad + \langle a_{\mathrm{p}}^{\beta}(t)a_{-\mathrm{q}'}^{\gamma'}\rangle\langle a_{\mathrm{q}}^{\gamma}(t)a_{-\mathrm{p}'}^{\beta'}\rangle \\ &= \delta_{\mathrm{p,p'}}\delta_{\mathrm{q,q'}}\langle a_{\mathrm{p}}^{\beta}(t)a_{-\mathrm{p}}^{\beta'}\rangle\langle a_{\mathrm{q}}^{\gamma}(t)a_{-\mathrm{q}}^{\gamma'}\rangle \\ &\quad + \delta_{\mathrm{p,q'}}\delta_{\mathrm{q,p'}}\langle a_{\mathrm{p}}^{\beta}(t)a_{-\mathrm{p}}^{\gamma'}\rangle\langle a_{\mathrm{q}}^{\gamma}(t)a_{-\mathrm{q}}^{\beta'}\rangle. \qquad (2.34)\end{aligned}$$

The Kronecker delta functions in the last expression result from the fact that the total wavevector in a correlation function must add up to zero if the motions in the liquid are homogeneous.

A second problem to be faced in carrying out the approximation scheme arises because of the Cartesian component labels attached to the autocorrelation functions, which means that the exponential in the integrand cannot immediately be replaced by an autocorrelation as is required.

In view of (2.26), which results from the incompressibility condition, the autocorrelation function must satisfy

$$\sum_{\alpha} k_{\alpha}\langle a_{\mathrm{k}}^{\alpha}(t)a_{-\mathrm{k}}^{\alpha'}\rangle = \sum_{\alpha'} k_{\alpha'}\langle a_{\mathrm{k}}^{\alpha}(t)a_{-\mathrm{k}}^{\alpha'}\rangle = 0, \qquad (2.26a)$$

ISBN 0-8053-6610-5, 0-8053-6611-3 (pbk.)

for all values of **k**. Moreover, the motions in the liquid are assumed to be isotropic, and therefore $\langle a_{\mathbf{k}}^{\alpha}(t)a_{-\mathbf{k}}^{\alpha'}\rangle$ does not change when the coordinate axes are arbitrarily rotated. This later requirement and (2.26a) are satisfied by taking the autocorrelation function to be of the form,

$$\langle a_{\mathbf{k}}^{\alpha}(t)a_{-\mathbf{k}}^{\alpha'}\rangle = D_{\mathbf{k}}^{\alpha\alpha'} g_k(t). \tag{2.35}$$

An equation for the new function $g_k(t)$ is then obtained from (2.33) by multiplying both sides by $D_{\mathbf{k}}^{\alpha'\alpha}$, summing over α and α', making the approximation to the fourth-order correlation function, and replacing the bare autocorrelation function with the full, interaction autocorrelation function, proportional to $g_k(t)$. The algebra involved in carrying out this program to obtain an equation for $g_k(t)$ is simplified by using the identity $\Sigma_{\alpha'} D_{\mathbf{k}}^{\alpha\alpha'}D_{\mathbf{k}}^{\alpha'\beta} = D_{\mathbf{k}}^{\alpha\beta}$ together with $\Sigma_{\alpha} D_{\mathbf{k}}^{\alpha\alpha} = (d-1)$, and $g_k(0) = 1$.

We find, after some straightforward algebra, that $g_k(t)$ satisfies the equation

$$\begin{aligned}(\partial_t + \nu_0 k^2)g_k(t) &= -\frac{k^2\chi}{(2\pi)^3}\int d\mathbf{p}\; T_{\mathbf{k}}(\mathbf{p}, \mathbf{k}-\mathbf{p}) \\ &\quad\times \int_0^t d\bar{t}\; g_p(\bar{t})g_{k-p}(\bar{t})g_k(t-\bar{t}) \\ &= -k^2\int_0^t d\bar{t}\; Q_k(\bar{t})g_k(t-\bar{t}),\end{aligned} \tag{2.36}$$

where the second equality defines the function $Q_k(t)$. The quantity $T_{\mathbf{k}}(\mathbf{p}, \mathbf{q})$ is`

$$\begin{aligned}4T_{\mathbf{k}}(\mathbf{p}, \mathbf{q}) &= (1-(\hat{\mathbf{k}}\cdot\hat{\mathbf{p}})^2)(1+(\hat{\mathbf{k}}\cdot\hat{\mathbf{q}})^2) + (1-(\hat{\mathbf{k}}\cdot\hat{\mathbf{q}})^2)(1+(\hat{\mathbf{k}}\cdot\hat{\mathbf{p}})^2) \\ &\quad + 2(\hat{\mathbf{k}}\cdot\hat{\mathbf{p}})(\hat{\mathbf{k}}\cdot\hat{\mathbf{q}})(\hat{\mathbf{p}}\cdot\hat{\mathbf{q}} - (\hat{\mathbf{k}}\cdot\hat{\mathbf{p}})(\hat{\mathbf{k}}\cdot\hat{\mathbf{q}})).\end{aligned} \tag{2.37}$$

Equation (2.36) represents an approximate, self-consistent equation for $g_k(t)$. At first sight it might appear too complicated to yield to analysis but we shall find that it can be solved in the limit of small **k** (long wavelengths).

A first step is to take advantage of the fact that the interaction term is in the form of a convolution integral. If the definition of $g_k(t)$ is modified so that it is zero for $t \leq 0$, then the Laplace transform

$$\tilde{g}_k(s) = \int_0^\infty dt\; g_k(t)\exp(-st) \tag{2.38}$$

satisfies the equation,

$$\{s + k^2(\nu_0 + \tilde{Q}_k(s))\}\tilde{g}_k(s) = g_k(t=0) = 1, \tag{2.39}$$

where $\tilde{Q}_k(s)$ is the Laplace transform of $Q_k(t)$ introduced in (2.36). The form of the result (2.39) suggests that we regard the nonlinear terms in the equation-of-motion as leading to a frequency and wavevector dependent

ISBN 0-8053-6610-5. 0-8053-6611-3 (pbk.)

kinematic viscosity. This concept of frequency and wavevector dependent transport coefficients in nonlinear systems is well established in the literature, and it stems, essentially, from pioneering work by Kubo (1957).

Before embarking on an analysis of (2.39) we pause to discuss the nature of the approximation we have made. While we have already made some comments on the nature of our approximation scheme, it is valuable to see how it is obtained from a perturbation expansion, the expansion being in terms of the nonlinear term in the equation-of-motion (2.27).

The use in statistical physics of various diagrammatic perturbation theory formalisms borrowed from quantum field theory is commonplace today, and there exist several excellent books on the subject [Abrikosov *et al.* (1963), and Fetter and Walecka (1971)]. There are, in fact, many methods of formulating perturbation schemes and the choice of method is, to some extent, one of personal taste. It would be out of place here to attempt even a brief review of the methods most frequently used; a comprehensive review would result in a tome.

The development of perturbation theories for the dynamic properties of classical systems has received its impetus from interest in the theory of turbulent motion in liquids and the incompressible Navier-Stokes liquid is the standard model for the study of the homogeneous isotropic turbulent state at intermediate wavevectors. Whereas an adequate description of the dynamic properties of condensed matter systems is often achieved with the summation of a limited class of diagrams, it is well established that a good theory of turbulence requires much more than this [Kraichnan (1972) and Monin and Yaglom (1975)]. There are some similarities between the turbulent state and the critical regime even though the two types of phenomena occur in different wavevector ranges. The nonlinear terms in the turbulent regime are such that a meaningful result is obtained only after a sophisticated renormalization of the perturbation expansion. In consequence, the formalism for diagrammatic perturbation expansions for properties of classical systems described by stochastic equations is in an advanced state, and the subject of continued research; see, for example, Garrido and San Miguel (1978).

Several authors have developed diagrammatic expansion schemes for the correlation function $g_k(t)$ [Deker and Haake (1975a) and Kawasaki (1976)], and their results can be expressed in a form similar to (2.36), namely,

$$(\partial_t + \nu_0 k^2) g_k(t) = - \int_0^t d\bar{t} \; g_k(t - \bar{t}) \Sigma_k(\bar{t}), \tag{2.40}$$

where $\Sigma_k(t)$ is identified as a collisional self-energy. The expansion of $\Sigma_k(t)$ is even in powers of $\chi^{1/2}$, and each diagram has a factor of k^2 associated with it, just as we found in our approximate equation-of-motion.

ISBN 0-8053-6610-5, 0-8053-6611-3 (pbk.)

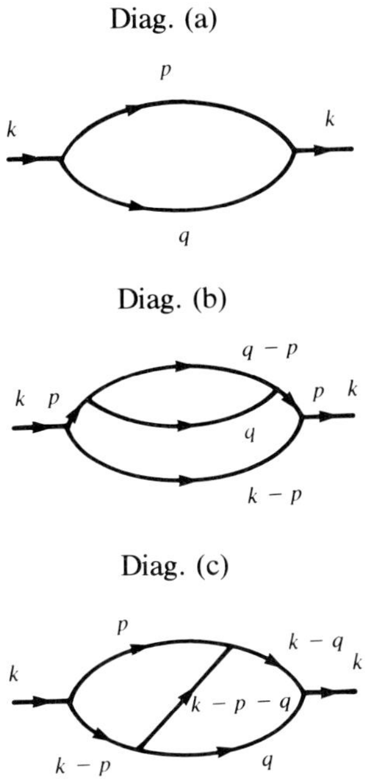

Fig. (2.2) The first three diagrams in the expansion of the collisional self-energy (2.40) in powers of $\chi^{1/2}$. Diagrams (a), (b) and (c) are of order χ and χ^2, respectively.

The first few diagrams for $\Sigma_k(t)$ are shown in Fig. (2.2) where each dot is a vertex and a line with a dot at each end is a bare propagator, $\exp(-\nu_0 k^2 t)$. The rules for constructing $\hat{\Sigma}_k(s)$ are readily appreciated with the aid of a couple of examples. Denoting the bare vertex function by $V_{\mathbf{k}}(\mathbf{p},\mathbf{q})$, Diag. (a) makes the following contribution to the Laplace transform of the collisional self-energy,

$$\int d\mathbf{p}\,\frac{V_{\mathbf{k}}(\mathbf{p},\mathbf{k}-\mathbf{p})V_{\mathbf{k}}^{*}(\mathbf{p},\mathbf{k}-\mathbf{p})}{s+\nu_0 p^2+\nu_0|\mathbf{k}-\mathbf{p}|^2} \tag{2.41}$$

and the contribution from Diag. (c) is,

$$\int d\mathbf{p}\int d\mathbf{q}\,\frac{V_{\mathbf{k}}(\mathbf{p},\mathbf{k}-\mathbf{p})}{(s+\nu_0 p^2+\nu_0|\mathbf{k}-\mathbf{p}|^2)} \times\frac{V_{\mathbf{k}-\mathbf{p}}(\mathbf{k}-\mathbf{p}-\mathbf{q},\mathbf{q})V_{\mathbf{k}-\mathbf{q}}^{*}(\mathbf{p},\mathbf{k}-\mathbf{p}-\mathbf{q})V_{\mathbf{k}}^{*}(\mathbf{k}-\mathbf{q},\mathbf{q})}{(s+\nu_0 p^2+\nu_0|\mathbf{k}-\mathbf{p}-\mathbf{q}|^2+\nu_0 q^2)(s+\nu_0|\mathbf{k}-\mathbf{q}|^2+\nu_0 q^2)} \tag{2.42}$$

The diagrams are seen to fall naturally into two classes. Diagrams (a) and (b) are called bubble diagrams. Diagram (c) differs from a bubble

ISBN 0-8053-6610-5, 0-8053-6611-3 (pbk.)

diagram by virtue of a propagator that links branches of a vertex, and such diagrams are called vertex corrections. Among the diagrams of order $(\chi^{1/2})^3$, there is one which has a dressed propagator as in Diag. (b) and a vertex correction of the type depicted in Diag. (c), and this diagram is classified as a vertex correction.

It is not difficult to verify that our approximation (2.36) is equal to the sum of all bubble diagrams for the collisional self-energy. To see this we consider the function

$$\int dt \exp(-st) \int d\mathbf{p} \,|\, V_{\mathbf{k}}(\mathbf{p}, \mathbf{k} - \mathbf{p})|^2 g_p(t) g_{k-p}(t) \tag{2.43}$$

which is, apart from a factor k^2, the function $\tilde{Q}_k(s)$ in (2.39). If we replace $g_p(t)$ and $g_{k-p}(t)$ in (2.43) by the free-propagators $\exp(-\nu_0 p^2 t)$ and $\exp(-\nu_0|\mathbf{k} - \mathbf{p}|^2 t)$, respectively, then we obtain the result (2.41) for Diag. (a), the lowest-order bubble diagram. To proceed further it is convenient to use the fact that the $g(t)$ functions are causal, so that (2.43) can be written $(\eta \to 0)$, apart from the momentum integral and the factor $|V|^2$,

$$\begin{aligned} &\left(\frac{1}{2\pi}\right)^2 \int_{-\infty}^{\infty} \frac{du\, du'}{(s - iu - iu' + \eta)} \tilde{g}_p(iu) \tilde{g}_{k-p}(iu') \\ &= \left(\frac{1}{2\pi}\right)^2 \int_{-\infty}^{\infty} du\, du' \{(s - iu - iu' + \eta)(iu + \nu_0 p^2 \\ &\quad + \tilde{\Sigma}_p(iu))(iu' + \nu_0 |\mathbf{k} - \mathbf{p}|^2 + \tilde{\Sigma}_{k-p}(iu'))\}^{-1}. \end{aligned} \tag{2.44}$$

We wish to verify that this expression generates all the bubble diagrams when iterated in the self-energy. For $\tilde{\Sigma}(s) = 0$ we recover (2.41). Expanding (2.44) to first-order in the self-energy terms, and using (2.41) for each self-energy, we find that (2.44) is the sum of the contributions from Diags. (a) and (b). Higher-order terms are then generated in an obvious fashion.

We conclude that our approximation (2.36) is equivalent in the diagrammatic expansion to summing all the bubble diagrams, i.e. $Q_k(t)$ is the sum of an infinite number of diagrams. Since there is no small parameter in the theory, this type of approximation is about as good as one might expect to be able to do. However, there is not much to guide us as to which class of diagrams to sum, and in some cases it might, for example, be most sensible to sum the ladder diagrams which include diagrams of the type (a) and (c), and excludes dressed propagators such as we see in Diag. (b).

As a prelude to investigating the hydrodynamic properties of the self-consistent equation for $Q_k(t)$ we calculate $Q_k(t)$ correct to first order in χ, i.e. we evaluate the contribution depicted by the diagram in Fig. (2.2a). The definition of $Q_k(t)$ given in (2.36) is,

$$Q_k(t) = \frac{\chi}{(2\pi)^3} \int d\mathbf{p}\; T_{\mathbf{k}}(\mathbf{p}, \mathbf{k} - \mathbf{p}) g_p(t) g_{k-p}(t), \tag{2.45}$$

ISBN 0-8053-6610-5, 0-8053-6611-3 (pbk.)

and the first-order result is obtained from this by replacing the autocorrelation functions by their noninteracting values, namely $g_p(t) = \exp(-\nu_0 p^2 t)$. In the limit of long wavelength, $T_{\mathbf{k}}(\mathbf{p}, \mathbf{k}-\mathbf{p})$ in (2.45) can be approximated by,

$$T_{\mathbf{k}}(\mathbf{p}, -\mathbf{p}) = \tfrac{1}{2} \sin^2 \theta(3 - 2 \sin^2 \theta), \tag{2.46}$$

where θ is the angle between $\mathbf{k}$ and $\mathbf{p}$. Using this result we find from (2.45),

$$\begin{aligned}\tilde{Q}_k(s) &= \int_0^\infty dt\, Q_k(t) \exp(-st) \\ &= \frac{\chi}{2(2\pi)^3} \int d\mathbf{p} \sin^2 \theta(3 - 2 \sin^2 \theta)/(s + \nu_0 p^2 + \nu_0 |\mathbf{k}-\mathbf{p}|^2).\end{aligned}$$

If we set $\mathbf{k} = 0$ in the integrand the angular integration can be completed, and it gives a factor $56\pi/15$. Hence, the long-wavelength limit of the first-order result for $\tilde{Q}_k(s)$ is,

$$\tilde{Q}_0(s) = \frac{7\chi}{30\pi^2} \int_0^\Lambda dp\, p^2\, (s + 2\nu_0 p^2)^{-1},$$

where Λ is the inverse of a typical molecular length. Notice that the integral is divergent in the limit $\Lambda s^{-1/2} \to \infty$. The divergent term can be extracted by rearranging the integrand into the following form,

$$\tilde{Q}_0(s) = \frac{7\chi}{60\pi^2 \nu_0} \int_0^\Lambda dp \left[1 - \frac{s}{2\nu_0} (p^2 + s/2\nu_0)^{-1} \right].$$

The second integral is finite for $\Lambda s^{-1/2} \to \infty$, and it is evaluated in this limit. We then obtain the following result for the first-order value of $\tilde{Q}_k(s)$ in the limit of long wavelengths,

$$\tilde{Q}_0(s) = \chi \left\{ \frac{7\Lambda}{60\pi^2 \nu_0} - \frac{7}{120\pi\nu_0} (s/2\nu_0)^{1/2} \right\}, \qquad s \to 0. \tag{2.47}$$

The first term of (2.47) is the first-order constant addition to the bare kinematic viscosity ν_0 due to the mixing of the velocity modes by the quadratic term in the Navier-Stokes equation. The observed kinematic viscosity is the sum of the bare viscosity and the total constant addition generated by the nonlinearity. Recall that for a Boltzmann distribution of the velocities the susceptibility χ is proportional to the temperature.

An interesting consequence of the result (2.47) is that the long-time behavior of $Q_0(t)$ must be of the form*

$$Q_0(t) \sim t^{-3/2}, \qquad t \to \infty.$$

* If for $t \to \infty$ the function $b(t) \sim t^{-\mu}$, where $\mu > 1$, then the Laplace transform $\tilde{b}(s) \sim \tilde{b}(0) + \pi s^{\mu-1}/\Gamma(\mu) \sin(\pi\mu)$ in the limit of small s, and $\Gamma(\mu)$ is the Gamma function.

ISBN 0-8053-6610-5, 0-8053-6611-3 (pbk.)

This is an example of what is usually referred to in the literature as a long-time tail contribution to a transport coefficient. Until quite recently it was generally assumed that transport coefficients decayed exponentially at long times, but such a decay is overridden, of course, by an inverse power-law behavior. The first evidence for long-time tails came, in fact, from computer simulation experiments on the dynamic properties of a gas of hard disks. Since then a substantial body of theoretical work has been published on the long-time behavior of transport coefficients, and the reader is referred to an excellent review by Pomeau and Résibois (1975).

Let us turn now to the investigation of the self-consistent equation for $\tilde{Q}_k(i\omega)$. From (2.36), (2.39) and (2.44) we find,

$$\begin{aligned}\tilde{Q}_k(i\omega) = \frac{\chi}{8\pi^2}\int_0^{\pi} d\theta \sin^3\theta(3 - 2\sin^2\theta)\int_0^{\Lambda} dp\, p^2 \int_{-\infty}^{\infty}\frac{du\, du'}{(2\pi)^2} \\ \times \{(i\omega - iu - iu' + \eta)[iu + \nu_0 p^2 + p^2\tilde{Q}_p(iu)] \\ \times [iu' + \nu_0|\mathbf{k}+\mathbf{p}|^2 + |\mathbf{k}+\mathbf{p}|^2\tilde{Q}_{k-p}(iu')]\}^{-1}. \qquad (2.48)\end{aligned}$$

We consider first the $\mathbf{k} \to 0$ limit of this expression since it is easier to handle than the $\omega \to 0$ limit. When we set $\mathbf{k} = 0$ in the right-hand side of (2.48) the angular integration can be completed immediately, and the factor is identical with the one which arises in the first-order calculation of $\tilde{Q}_k(s)$. Following the angular integration, it is expedient to make the change of variables $p \to (\omega/\nu_0)^{1/2}p = \xi p$ and $u \to \omega u$, where the new integration variables are dimensionless. We then have,

$$\begin{aligned}\tilde{Q}_0(i\omega) = \frac{7\chi}{30\pi^2\nu_0}\xi\int_0^{\Lambda/\xi} dp\, p^2\int_{-\infty}^{\infty}\frac{du\, du'}{(2\pi)^2} \\ \times\left\{(i\omega - iu - iu' + \eta)\left[iu + p^2 + p^2\frac{1}{\nu_0}\tilde{Q}_{\xi p}(iu\omega)\right]\right. \\ \left.\times\left[iu' + p^2 + p^2\frac{1}{\nu_0}\tilde{Q}_{\xi p}(iu'\omega)\right]\right\}^{-1}. \qquad (2.49)\end{aligned}$$

In the limit of very small ω we anticipate that,

$$\tilde{Q}_{\xi p}(iu\nu_0\xi^2) = b\nu_0, \qquad \xi \to 0, \qquad (2.50)$$

where b is a dimensionless constant independent of $\mathbf{p}$ and ω. On using this result in (2.49) the calculation of $\tilde{Q}_0(i\omega)$ reduces to the same calculation as we performed to get the corresponding first-order result apart from the change $\nu_0 \to \nu_0(1 + b)$. We find that, for long wavelengths and small

ISBN 0-8053-6610-5, 0-8053-6611-3 (pbk.)

frequencies,

$$\tilde{Q}_0(i\omega) = \chi\left\{\frac{7\Lambda}{60\pi^2\nu_0(1+b)} - \frac{7}{120\pi\nu_0(1+b)}(i\omega/2\nu_0(1+b))^{1/2}\right\}, \qquad \omega \to 0, \tag{2.51}$$

from which it follows that b satisfies the equation

$$b(1+b) = (7\chi\Lambda/60\pi^2\nu_0^2). \tag{2.52}$$

Evidently, $\nu_0(1+b)$ is the measured kinematic viscosity at the level of approximation to which we are working. The long-time behavior of $Q_0(t)$ is the same as we found from the first-order calculation, (2.47).

The calculation of $\tilde{Q}_k(0)$ from the self-consistent equation (2.48) is slightly more involved than the preceding calculation of $\tilde{Q}_0(i\omega)$. Making the change of variables $p \to kp$ and $u \to k^2\nu_0 u$, and setting $\omega = 0$ we find, in the long-wavelength limit,

$$\begin{aligned}\tilde{Q}_k(0) &= \frac{k\chi}{8\pi^2\nu_0(1+b)}\int_0^\pi d\theta \sin^3\theta(3 - 2\sin^2\theta) \\ &\quad \times \int_0^{\Lambda/k} dp\, p^2(p + |\mathbf{1} - \mathbf{p}|^2)^{-1} \\ &= \frac{\chi}{8\pi^2\nu_0(1+b)}\int_{-1}^{1} d\mu\,(1 + \mu^2 - 2\mu^4) \\ &\quad \times \left[\tfrac{1}{2}\Lambda - \frac{k\pi}{4}\frac{(1-\mu^2)}{\sqrt{(2-\mu^2)}}\right] \\ &= \nu_0 b - \frac{k\chi}{24\nu_0(1+b)}\left(\frac{9\pi}{16} - 1\right).\end{aligned} \tag{2.53}$$

Notice that the two contributions to $\tilde{Q}_k(0)$ have opposite signs, and so the wavevector dependent term partially negates the constant contribution to the kinematic viscosity.

Assuming a Boltzmann distribution of the velocities we find $\chi = 27.6\ 10^{-16}$ cm^5 sec^{-2} for water at 20°C. The observed kinematic viscosity of water at 20°C is 10^{-2} cm^2 sec^{-1}. Using these values the frequency dependent contribution to the quantity ($\tilde{Q}_0(i\omega)$/observed kinematic viscosity) is, $-(i\ 13.2\omega\ 10^{-24})^{1/2}$ with ω in units of sec^{-1}. Hence, for $\omega = 10^9$ sec^{-1}, a typical frequency shift observed in light scattering experiments, the frequency dependent term in (2.51) is a very small contribution. A similar calculation shows that the wavevector dependent contribution to the quan-

ISBN 0-8053-6610-5, 0-8053-6611-3 (pbk.)

tity ($\tilde{Q}_k(0)$/observed kinematic viscosity) is, $-0.28k\ 10^{-4}$, with k in Å^{-1}, for water at 20°C. With $k = 10^{-2}$ Å^{-1}, the wavevector dependent contribution to (2.53) is a factor $\sim 10^{-7}$ smaller than the observed kinematic viscosity.

The type of approximation scheme developed in this section can be adapted to study the hydrodynamic properties of stochastic systems more complicated than the incompressible Navier-Stokes liquid. An example where the nonlinearity in the equation-of-motion is cubic, rather than quadratic, is studied in the last section of this chapter. Another complicating feature that can arise is when the stochastic equation contains more than one variable. A simple example of this type is afforded by the equation for the local temperature in a liquid, which is studied in Appendix 2A. Under certain conditions, called forced convection, temperature inhomogeneities simply move with the liquid, at the same time being smoothed out under the influence of molecular thermal conductivity. In consequence, the stochastic equation for the local temperature contains a coupling with the velocity field.

2.4 MODEL EQUATIONS

In the course of discussing the two models introduced in preceding sections we have come across several features which are general properties of stochastic models. One purpose of the present section is to draw attention to these properties, and to put the two models we have discussed into a more general framework. A second purpose is to discuss the construction of stochastic equations to describe physical processes; a nice review is given by Fox (1978).

Let us denote the dynamical variables by φ_i where the index i might, for example, label the various components of φ or be a wavevector. The simple Langevin equation discussed in Sec. 2.1 assumes that φ_i satisfies an equation-of-motion of the form

$$\partial_t \varphi_i = -(L_i/\chi_i)\varphi_i + f_i \tag{2.54}$$

where L_i are the transport coefficients, χ_i is the static susceptibility for φ_i, and f_i is a Gaussian-distributed random force that satisfies,

$$\langle f_i(t) f_j \rangle = 2L_i \delta_{i,j} \delta(t). \tag{2.55}$$

Comparing (2.54) and (2.55) with (2.1) and (2.5), respectively, we identify L and χ for the case when φ is the velocity of a particle undergoing Brownian motion as

$$L = \gamma/M\beta, \qquad \chi = 1/M\beta.$$

For the case of the incompressible Navier-Stokes liquid discussed in Sec. 2.3, χ is again proportional to temperature and inversely proportional to the mass. In the more general case when φ is not the velocity, and we

ISBN 0-8053-6610-5, 0-8053-6611-3 (pbk.)

can no longer use the equipartition theorem to evaluate the equal time autocorrelation function and hence χ, Landau and Lifshitz (1959) show that we then need to know how the rate of production of entropy is related to φ. A complete example of this argument is afforded by their treatment of the linearized hydrodynamic equations for a compressible liquid, Landau and Lifshitz (1963). If we compare (2.55) with the corresponding result for the Navier-Stokes liquid, Eq. (2.30), we see that, for this latter case, the transport coefficient is wavevector dependent, namely, $L \equiv L_{\mathbf{k}} = k^2\nu_0$. This feature is a consequence of the fact that the velocity field is a conserved quantity, i.e. it satisfies (2.12). Our present notation differs from the one used in the discussion of the diffusion constant as the integral of the velocity autocorrelation function, given in Sec. 2.2, in that the transport coefficient now contains the factor of wavevector squared whereas in Sec. 2.2 we choose to separate out this factor.

An equation-of-motion of the form (2.54) is compatible with a Gaussian equilibrium distribution function for the variables φ_i. The distribution function is usually written in terms of a free-energy function F such that

$$P_0 = \exp(-F). \tag{2.56}$$

A more general form for the equation-of-motion is then

$$\partial_t \varphi_i = -L_i \partial_{\varphi_i} F + f_i. \tag{2.57}$$

For a Gaussian distribution

$$F = \tfrac{1}{2} \sum_j \varphi_j^2 / \chi_j, \tag{2.58}$$

and we then recover (2.54) from (2.57). We shall later show that (2.57) generates the same static correlation functions as given by the equilibrium distribution function (2.56). For the moment we continue to examine features of the equation-of-motion.

The form of the free-energy is frequently derived by considering the change to it caused by the application of an external field. If $\mu(\mathbf{r})$ is the field conjugate to $\varphi(\mathbf{r})$ then the perturbation is described by a Hamiltonian density

$$\mathscr{H}_1 = -\int d\mathbf{r}\, \mu(\mathbf{r})\varphi(\mathbf{r}). \tag{2.59}$$

The change in $\varphi(\mathbf{r})$ due to a change in $\mu(\mathbf{r})$ is obtained from (1.62), taking into account that the variables φ and μ are classical,

$$\delta\langle\varphi(\mathbf{r})\rangle = \beta \int d\mathbf{r}'\, \delta\mu(\mathbf{r}')\langle\{\varphi(\mathbf{r}) - \langle\varphi(\mathbf{r})\rangle\}\,\{\varphi(\mathbf{r}') - \langle\varphi(\mathbf{r}')\rangle\}\rangle, \tag{2.60}$$

where $\delta\mu(\mathbf{r})$ is the change in $\mu(\mathbf{r})$, and the average value on the right-hand side of (2.60) is taken with respect to the unperturbed system. Taking $\delta\mu(\mathbf{r})$

ISBN 0-8053-6610-5, 0-8053-6611-3 (pbk.)

in (2.60) to be a constant $\delta\mu$,

$$\delta\varphi = \beta\delta\mu \int d\mathbf{r}' \langle \delta\varphi(\mathbf{r} = 0)\delta\varphi(\mathbf{r}')\rangle$$

from which it follows that

$$\partial_\mu \varphi = \beta \int d\mathbf{r} \langle \delta\varphi(0)\delta\varphi(\mathbf{r})\rangle, \tag{2.61}$$

where it is understood that the partial derivative is evaluated with all other relevant variables kept constant. For example, if φ is the mass density ρ of a liquid, μ is the chemical potential, and

$$(\partial\rho/\partial\mu)_{T,\Omega} = \rho(\partial\rho/\partial p)_T = \rho^2/B_T,$$

where p is the pressure and B_T is the isothermal bulk modulus. Equation (2.61) is important because it relates the thermodynamic derivative $(\partial\varphi/\partial\mu)$ to the integral over all space of the mean-square fluctuation of φ. Returning to the case where φ is identified with the mass density, we deduce that this integral diverges at the liquid-gas phase transition since $B_T \to 0$ at the transition.

Regarding the free-energy as a function of φ the deviation in the free-energy is

$$F = \frac{\beta}{\Omega} \int d\mathbf{r} \left[a_0\delta\varphi(\mathbf{r}) + \tfrac{1}{2}a_1\delta\varphi(\mathbf{r})^2 + \cdots - \delta\mu(\mathbf{r})\delta\varphi(\mathbf{r})\right].$$

The first term in the Taylor expansion when integrated over all space evidently vanishes since φ is conserved. The coefficient of the second term in the expansion is, by definition, $a_1 = (1/\beta)(\partial^2 F/\partial\varphi^2)$. For the example when φ is the mass density it is easy to show that $a_1 = B_T/\rho^2\beta$, and we conclude from this that a_1 is positive and vanishes at the critical point.

The terms included in the Taylor expansion of the free-energy are not the only ones that contribute. As a consequence of the perturbation the system is not homogeneous and accordingly we must include spatial derivatives of φ. If the system is isotropic the lowest-order term of this kind that can enter in the expansion is proportional to $(\boldsymbol{\nabla}\varphi)^2$. The coefficient of this term is taken to be a constant $\frac{1}{2}m_0^2$. For ease of notation we shall now write φ in place of $\delta\varphi$ in the free-energy, and obtain

$$F = \frac{\beta}{\Omega} \int d\mathbf{r} \left[\tfrac{1}{2}a_1\varphi^2(\mathbf{r}) + \tfrac{1}{2}m_0^2(\boldsymbol{\nabla}\varphi(\mathbf{r}))^2 + \cdots - \mu(\mathbf{r})\varphi(\mathbf{r})\right]. \tag{2.62}$$

The next term in the Taylor expansion in (2.62) is φ^4 for most models of interest, and the resulting form for the free-energy is usually called the Landau-Ginzburg free-energy.

An equation for $\varphi(\mathbf{r})$ is obtained from (2.62) by requiring that the free-energy be a minimum with respect to φ. With the terms shown explicitly in

ISBN 0-8053-6610-5, 0-8053-6611-3 (pbk.)

(2.62) this equation is

$$(a_1 - m_0^2\nabla^2)\varphi(\mathbf{r}) = \mu(\mathbf{r}). \tag{2.63}$$

If we operate on (2.60) with the operator that occurs on the left-hand side of (2.63) then

$$\int d\mathbf{r}'\ \mu(\mathbf{r}')\{\beta(a_1 - m_0^2\nabla_r^2)\langle\varphi(\mathbf{r})\varphi(\mathbf{r}')\rangle - \delta(\mathbf{r}-\mathbf{r}')\} = 0.$$

Because $\mu(\mathbf{r})$ is arbitrary

$$(\nabla^2 - a_1/m_0^2)\langle\varphi(\mathbf{r})\varphi(0)\rangle = -(1/m_0^2\beta)\delta(\mathbf{r}) \tag{2.64}$$

and making use of the identity $\nabla^2(1/r) = -4\pi\delta(\mathbf{r})$ we readily verify that the solution of (2.64) that is zero for $r \to \infty$ is the Ornstein-Zernike form

$$\langle\varphi(\mathbf{r})\varphi(0)\rangle = \frac{1}{4\pi m_0^2\beta}\exp(-r/\xi)/r \tag{2.65}$$

where the correlation length

$$\xi = (a_1/m_0^2)^{-1/2}. \tag{2.66}$$

The final topic in the present discussion of the construction of a free-energy is to write it in terms of the Fourier components

$$\varphi_{\mathbf{k}} = \Omega^{-1}\int d\mathbf{r}\ \varphi(\mathbf{r})\exp(i\mathbf{k}\cdot\mathbf{r}). \tag{2.67}$$

Keeping in (2.62) terms that are quadratic in φ,

$$\begin{aligned} F &= \tfrac{1}{2}\sum_{\mathbf{k}}\{\beta(a_1 + m_0^2k^2)|\varphi_{\mathbf{k}}|^2 - 2\beta\varphi_{\mathbf{k}}\mu_{-\mathbf{k}}\} \\ &= \tfrac{1}{2}\sum_{\mathbf{k}}\{\chi_k^{-1}|\varphi_{\mathbf{k}}|^2 - 2\beta\varphi_{\mathbf{k}}\mu_{-\mathbf{k}}\} \end{aligned} \tag{2.68}$$

where the second equality defines the wavevector dependent susceptibility, χ_k. From (2.65),

$$\begin{aligned} \langle|\varphi_{\mathbf{k}}|^2\rangle &= \Omega^{-1}\int d\mathbf{r}\ \langle\varphi(\mathbf{r})\varphi(0)\rangle \\ &= (\Omega m_0^2\beta)^{-1}(k^2 + \xi^{-2})^{-1} = \chi_k/\Omega. \end{aligned} \tag{2.69}$$

The equation-of-motion satisfied by $\varphi_{\mathbf{k}}$ is obtained from (2.57) using (2.68) for the free-energy,

$$\begin{aligned} \partial_t\varphi_{\mathbf{k}} &= -L_k\partial_{\varphi_{\mathbf{k}}}F + f_{\mathbf{k}} \\ &= -(L_k/\chi_k)\varphi_{\mathbf{k}} + \beta\mu_{\mathbf{k}} + f_{\mathbf{k}}. \end{aligned} \tag{2.70}$$

If the free-energy contains terms of higher order in φ than quadratic, e.g. the Landau-Ginzburg free-energy, then these higher-order terms will generate nonlinear terms in the equation-of-motion.

ISBN 0-8053-6610-5, 0-8053-6611-3 (pbk.)

A second source of nonlinear terms are, so-called, mode-couplings. We have already come across such terms in the Navier-Stokes liquid where they arise from purely kinematic considerations. A second example is afforded by the equation-of-motion for the local spin density in a ferromagnet. The local magnetic field **H** gives rise to a torque term in the equation-of-motion, in addition to the diffusive term. Denoting the spin density by **S** the equation-of-motion in real space is

$$\partial_t \mathbf{S} = \lambda_0 \mathbf{S} \times \mathbf{H} - \nu_0 \nabla^2 \mathbf{H} + \mathbf{f}. \tag{2.71}$$

Here λ_0 is the strength of the mode-coupling term, and

$$\mathbf{H} = \mathbf{H}_0 - (\partial F/\partial \mathbf{S}), \tag{2.72}$$

where $\mathbf{H}_0$ is a steady external field.

If we include mode-coupling terms in the equation-of-motion by the addition of a streaming velocity $V(\varphi)$ then our prototype equation-of-motion becomes

$$\partial_t \varphi_i = V_i(\varphi) - L_i \partial_{\varphi_i} F(\varphi) + f_i. \tag{2.73}$$

Equation (2.73) has been derived within certain assumptions by several authors using the projection operator technique discussed in Chap. III. From the derivation we obtain an explicit formula for the streaming velocity [Mori and Fujisaka (1973) and Kawasaki (1976)]

$$V_i(\varphi) = \lambda_0 \sum_j P_0^{-1} \partial_{\varphi_j} \{P_0 Q_{ij}\}, \tag{2.74}$$

where Q_{ij} are constructed from Poisson brackets or commutators of φ_i, and $Q_{ij} = -Q_{ji}$. An important property of the streaming velocity is that the probability current VP_0 is divergence-free in φ-space, i.e.

$$\sum_i \partial_{\varphi_i} \{V_i(\varphi) P_0\} = 0. \tag{2.75}$$

Because a divergence-free current does not change the distribution function the inclusion of the streaming velocity V in the equation-of-motion does not change equilibrium properties determined by (2.56), e.g. the wavevector dependent susceptibility does not depend on the mode-coupling strength parameter λ_0.

The main assumptions made in derivations of (2.73) are (a) dynamic processes which contribute to the transport coefficients L_i occur on a time-scale which is very short compared to the characteristic time-scale for the slow variables φ_i, (b) L_i do not depend on the variables φ_i, and (c) the random force has a Gaussian distribution.

We shall now demonstrate that the equation-of-motion (2.73) leads to the same equilibrium correlation functions as are obtained from the distribution function (2.56); very detailed discussions are given by Graham (1973) and Ma and Mazenko (1975). To this end we introduce new variables a_i

ISBN 0-8053-6610-5, 0-8053-6611-3 (pbk.)

and a new function

$$P(t) = \prod_j \delta(a_j - \varphi_j(t)). \tag{2.76}$$

The virtue of introducing these new quantities is apparent when we observe that

$$\langle \varphi_1(t)\varphi_2(t)\cdots\varphi_j(t)\rangle = \int \prod_i da_i a_1 a_2 \cdots a_j \langle P(t)\rangle. \tag{2.77}$$

In consequence, $\langle P(t)\rangle$ is the equilibrium distribution function associated with the variables φ_j, and we wish to show that this function is the same as (2.56). The equation-of-motion for $P(t)$ is

$$\begin{aligned}\partial_t P(t) &= -\sum_j \partial_{a_j} P(t)\partial_t \varphi_j(t) \\ &= -\sum_j \partial_{a_j} P(t)\{V_j(a) - L_j \partial_{a_j} F(a) + f_j\}.\end{aligned} \tag{2.78}$$

The equation-of-motion for $\langle P(t)\rangle$ is obtained from (2.78) together with the result, valid for Gaussian random forces [Novikov (1965)]

$$\langle f_i(t)P(t)\rangle = -L_i \partial_{a_i}\langle P(t)\rangle \tag{2.79}$$

and the result is

$$\partial_t \langle P(t)\rangle = \mathscr{L}\langle P(t)\rangle \tag{2.80}$$

where the Fokker-Planck operator [Résibois and de Leener (1977)]

$$\mathscr{L} = -\sum_j \partial_{a_j}[V_j - L_j(\partial_{a_j} + \partial_{a_j} F)]. \tag{2.81}$$

Using the divergence condition on the streaming velocity (2.75) it is readily verified that

$$\mathscr{L} P_0(a) = \mathscr{L} \exp(-F(a)) = 0 \tag{2.82}$$

from which we conclude that the equilibrium distribution function (2.56) is a static solution of the Fokker-Planck equation. An additional important result, which follows from (2.78) is

$$\langle \varphi_i \varphi_j(t)\rangle = \int \prod_{j'} da_{j'} a_i \exp(\mathscr{L} t) a_j P_0(a). \tag{2.83}$$

This shows that $\mathscr{L}$ is the time evolution operator for the variables a_j.

To complete our present discussion we study linear response theory for stochastic systems. An external perturbation of strength $h(t)$ is assumed to couple to the variables φ resulting in a change in the free-energy function of the form, cf. Eq. (2.62),

$$F \rightarrow F - \sum_j h_j(t)\varphi_j. \tag{2.84}$$

ISBN 0-8053-6610-5, 0-8053-6611-3 (pbk.)

This change modifies the streaming velocity (2.74), and the Fokker-Planck operator (2.81), is replaced by

$$\mathscr{L} \to -\sum_j \partial_{a_j}[V_j - L_j(\partial_{a_j} + \partial_{a_j}F)] - \sum_{ij} h_j(t)\partial_{a_i}\{\delta_{ij}L_i + \lambda_0 Q_{ij}\}. \tag{2.85}$$

Treating the modification to $\mathscr{L}$ as a small perturbation, the first-order change in the average value of φ_i induced by $h(t)$ is [compare with Eq. (1.42)]

$$\sum_j \int_{-\infty}^{t} d\bar{t}\, h_j(\bar{t})K_{ij}(t-\bar{t}), \tag{2.86}$$

where, for $t > 0$, the response function is [Garrido and San Miguel (1978)]

$$\begin{aligned} K_{ij}(t) &= -\int \prod_{j'} da_{j'}a_i \exp(\mathscr{L}t) \sum_{i'} \partial_{a_{i'}}\{\delta_{i'j}L_j + \lambda_0 Q_{i'j}\}P_0(a) \\ &= -\int \prod_{j'} da_{j'}a_i \exp(\mathscr{L}t)\mathscr{L}a_jP_0(a) \\ &\equiv -\partial_t\langle\varphi_i\varphi_j(t)\rangle. \end{aligned} \tag{2.87}$$

The second equality in (2.87) follows from the definition of the streaming velocity (2.74), and the use of the divergence-free condition on the probability current. We have,

$$\sum_i \partial_{a_i}\{L_i\delta_{i,j} + \lambda_0 Q_{ij}\}P_0 = L_j\partial_{a_j}P_0 - P_0V_j \tag{2.88}$$

and using the definition of the Fokker-Planck operator (2.81)

$$\mathscr{L}a_jP_0 = -\sum_i \{\delta_{i,j}V_jP_0 + a_j\partial_{a_i}(V_iP_0)\} + L_j\partial_{a_j}P_0 \tag{2.89}$$

which, together with (2.88), gives the identity required to obtain the second form in (2.87) for the response function $K_{ij}(t)$. These same identities can be used to prove that $K_{ij}(t=0)$ is a constant independent of the nature of the nonlinear terms in the equation-of-motion. From (2.87), (2.88) and (2.74) we have

$$\begin{aligned} K_{ij}(0) &= -\int \prod_{j'} da_{j'}a_i(L_j\partial_{a_j}P_0 + \lambda_0 \sum_{i'} \partial_{a_{i'}}\{Q_{i'j}P_0\}) \\ &= \int \prod_{j'} da_{j'}(\delta_{ij}L_jP_0 + \lambda_0 Q_{ij}P_0) = \delta_{ij}L_j \end{aligned} \tag{2.90}$$

where the last line follows from the normalization condition on P_0, and the definition of Q_{ij} in terms of a Poisson bracket or commutator of the variables φ_i, φ_j. Notice that if the streaming velocity is zero, it follows from (2.87)

ISBN 0-8053-6610-5, 0-8053-6611-3 (pbk.)

and (2.89) that the response function

$$K_{ij}(t) = -L_j\langle \varphi_i \exp(\mathscr{L}t)\partial_{\varphi_j}\rangle$$
$$\equiv -L_j\langle \varphi_i(t)\partial_{\varphi_j}\rangle \quad \text{for } V_j(\varphi) = 0 \tag{2.91}$$

that is to say, the response function is the correlation function formed from the variables $\varphi_i(t)$ and ∂_{φ_j}. Because these variables satisfy the commutation relation

$$[\varphi_i, \partial_{\varphi_j}] = -\delta_{ij} \tag{2.92}$$

they can be used to construct perturbation theories for the response function by analogy with perturbation theories for quantum field theory, e.g. Martin *et al.* (1973). Alternatively, we can introduce Bose operators α_i, α_i^+ which satisfy

$$[\alpha_i, \alpha_j^+] = \delta_{ij} \tag{2.93a}$$

all other commutators being zero, and then write,

$$\varphi_i = \alpha_i + \alpha_i^+ \text{ and, } \partial_{\varphi_i} = \tfrac{1}{2}(\alpha_i - \alpha_i^+). \tag{2.93b}$$

Zwanzig (1972) exploits this transformation to set up a perturbation theory for the velocity autocorrelation function for the incompressible Navier-Stokes liquid. For this case the free-energy is quadratic in φ, and the transformation (2.93) reduces the part of $\mathscr{L}$ proportional to L to a harmonic oscillator form and the streaming velocity is then treated as a perturbation within the harmonic oscillator states.

2.5 TIME-DEPENDENT LANDAU-GINZBURG MODEL

As a final topic in this chapter we consider the dynamic properties of a system for which the equation-of-motion is

$$\partial_t \varphi_{\mathbf{k}} = -L_k \partial_{\varphi_{\mathbf{k}}^*} F + f_k \tag{2.94}$$

and

$$F = \frac{1}{\Omega}\int d\mathbf{r}\left\{\tfrac{1}{2}m_0^2\varphi^2(\mathbf{r}) + \tfrac{1}{2}(\nabla\varphi(\mathbf{r}))^2 + \frac{1}{4!\Omega}\, g_0\varphi^4(\mathbf{r})\right\}$$
$$= \sum_{\mathbf{k}} \tfrac{1}{2}(k^2 + m_0^2)|\varphi_{\mathbf{k}}|^2 + \frac{1}{4!\Omega}\, g_0 \sum_{1234} \delta_{1+3,2+4}\varphi_1\varphi_2^*\varphi_3\varphi_4^* \tag{2.95}$$

where L_k, m_0 and g_0 are parameters. In (2.95) we have introduced the shorthand notation for wavevectors, $\mathbf{k}_1 = 1$, etc. Using (2.95) the equation-

ISBN 0-8053-6610-5, 0-8053-6611-3 (pbk.)

of-motion is, explicitly,

$$\partial_t \varphi_k = -L_k(k^2 + m_0^2)\varphi_k - \frac{L_k}{3!\Omega} g_0 \sum_{234} \delta_{k+3,2+4} \varphi_2 \varphi_3^* \varphi_4. \qquad (2.96)$$

In constructing an approximate equation-of-motion for the correlation function

$$G_k(t) = \langle \varphi_k(t)\varphi_k^* \rangle, \qquad G_k(0) = \chi_k \qquad (2.97)$$

we follow the prescription adopted in Sec. 2.3. This requires us to formally integrate (2.96), and use the result to substitute for $\varphi_k^*(-t)$ in the interaction term in the equation-of-motion for $G_k(t)$. An important difference between the present problem and the problem discussed in Sec. 2.3 is that the interaction term is now a sixth-order correlation function and its reduction to products of G's involves two levels of decomposition. This extra level of decomposition obviously increases the error involved in replacing the correlation function by products compared to the case in Sec. 2.3. In the latter case we argued that the error introduced in decomposing the correlation function was partially offset by replacing the bare propagator by a full propagator. To achieve a comparable approximation for the sixth-order correlation function we must replace the bare propagator by a function of higher order than the full propagator. The time derivative of the full propagator is the next higher-order correlation function to the propagator so we should achieve the same level of approximation in our equation for $G_k(t)$ as we did in Sec. 2.3 if we decompose the sixth-order correlation function in the interaction term into a triple product of G's and, at the same time, replace the bare propagator by $\dot{G}_k(t)$. The decomposition follows the same prescription that we gave in Sec. 2.3, i.e. it is achieved by making all possible pairings from the sixth-order correlation function (the pairs of variables having different time arguments) and dividing the result by 3!. This prescription leads to the equation

$$\begin{aligned}\partial_t G_k(t) &\equiv \dot{G}_k(t) \\ &= -\omega_k G_k(t) + \frac{1}{6\dot{G}_k(0)}\left(\frac{L_k g_0}{\Omega}\right)^2 \int_0^t d\bar{t}\, \dot{G}_k(t-\bar{t}) \\ &\quad \times \sum_{234} \delta_{k+3,2+4} G_2(\bar{t}) G_3(\bar{t}) G_4(\bar{t}) \end{aligned} \qquad (2.98)$$

where

$$\omega_k = L_k(k^2 + m_0^2) + \frac{L_k g_0}{6\chi_k \Omega} \sum_{234} \delta_{k+3,2+4} \langle \varphi_2 \varphi_3^* \varphi_4 \varphi_k^* \rangle. \qquad (2.99)$$

The form of the first term in (2.98), involving the frequency ω_k, follows directly from the equation-of-motion (2.97). We can take advantage of the

ISBN 0-8053-6610-5, 0-8053-6611-3 (pbk.)

results (2.90) and (2.91) to show that

$$\dot{G}_k(0) = -L_k, \quad \text{or} \quad \omega_k \chi_k = L_k \tag{2.100}$$

which means that

$$\chi_k(k^2 + m_0^2) + \frac{1}{6\Omega} g_0 \sum_{234} \delta_{k+3,2+4} \langle \varphi_2 \varphi_3^* \varphi_4 \varphi_k^* \rangle = 1. \tag{2.101}$$

Formally, at least, the susceptibility χ_k is obtained from (2.101) by calculating the fourth-order equilibrium correlation function from the equilibrium distribution function (2.56) with the free-energy given by (2.84). However, it might happen that the equilibrium correlation function can be calculated to a good approximation by some other method, or at least its wavevector dependence can be obtained. In this instance, equilibrium correlation functions can be taken to be known quantities, as is the case for L_k.

We define a collisional self-energy $\Sigma_k(t)$ through the equation,

$$\partial_t G_k(t) = -\frac{L_k}{\chi_k} G_k(t) - \frac{1}{L_k} \int_0^t d\bar{t}\, \dot{G}_k(t - \bar{t}) \Sigma_k(\bar{t}) \tag{2.102}$$

and comparing this equation with (2.98) our approximation for the self-energy is

$$\Sigma_k(t) \simeq \Sigma_k^{(1)}(t) = \frac{1}{6} \left(\frac{L_k g_0}{\Omega} \right)^2 \sum_{234} \delta_{k+3,2+4} G_2(t) G_3(t) G_4(t). \tag{2.103}$$

If the definition of $G_k(t)$ is modified to make it a causal function then its Laplace transform $\tilde{G}_k(s)$ satisfies the equation

$$\tilde{G}_k(s) = \chi_k(1 + L_k^{-1} \tilde{\Sigma}_k(s)) \left\{ s + (L_k/\chi_k) + \frac{s}{L_k} \tilde{\Sigma}_k(s) \right\}^{-1} \tag{2.104}$$

where $\tilde{\Sigma}_k(s)$ is the Laplace transform of the self-energy, and the corresponding equation for the response function $K_k(t) = -\partial_t G_k(t)$ is,

$$\tilde{K}_k(s) = L_k \left\{ s + (L_k/\chi_k) + \frac{s}{L_k} \tilde{\Sigma}_k(s) \right\}^{-1}. \tag{2.105}$$

If the equilibrium correlation functions are known then there is at least one improvement that we can make to (2.103). Differentiating (2.102) with respect to t and setting $t = 0$,

$$\ddot{G}_k(0) = (L_k^2/\chi_k) + \Sigma_k(t = 0). \tag{2.106}$$

We build this result into our approximation for the self-energy by taking

$$\Sigma_k(t) \simeq (\ddot{G}_k(0) - L_k^2/\chi_k) \Sigma_k^{(1)}(t) / \Sigma_k^{(1)}(t = 0). \tag{2.107}$$

ISBN 0-8053-6610-5, 0-8053-6611-3 (pbk.)

In this prescription we have built exact equilibrium quantities into the equation for $G_k(t)$, and in doing so we eliminate the explicit dependence on the coupling constant g_0.

A diagrammatic expansion for $G_k(t)$ has been developed by Deker and Haake (1975b), and it is instructive to compare their results with our approximate treatment. Their equation for $G_k(t)$ is of the same form as (2.98) with ω_k given by (2.100). The expansion of the collisional self-energy in terms of $G(t)$ is shown in Fig. (2.3); note that the expansion is in terms of $G(t)$ and $\dot{G}(t)$ and not a bare propagator as is this case in Fig. (2.2) for the incompressible Navier-Stokes liquid. The first diagram corresponds exactly to the approximation (2.98), and the precise form of the vertex appearing in the expansion can be extracted from (2.98). The next diagram in the expansion of $\Sigma_k(t)$ in terms of the number of vertices includes the response function as well as G_k itself.

An important additional result obtained from perturbation theory relates the self-energy at $t = 0$ to the fourth-order correlation function in ω_k, namely,

$$\Sigma_k(t=0) = \frac{1}{2}\frac{L_k^2}{\Omega} g_0 \left\{ -\sum_{p} \chi_p + \frac{1}{3\chi_k}\sum_{234} \delta_{k+2,3+4}\langle \varphi_2\varphi_3^*\varphi_4\varphi_k^*\rangle \right\}. \qquad (2.108)$$

The lowest-order approximation for the static fourth-order correlation func-

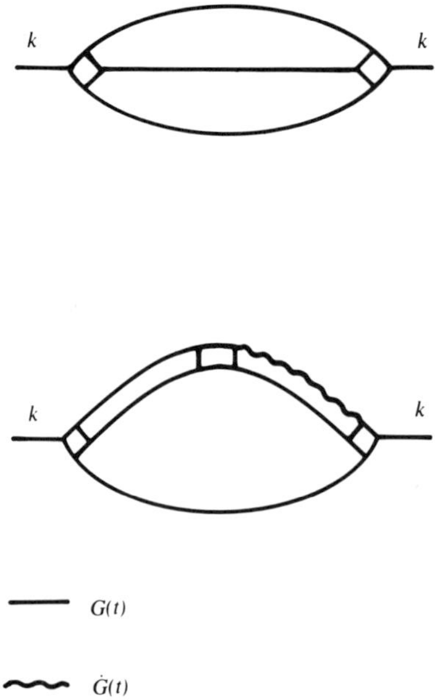

Fig. (2.3) The first two terms in the expansion of the self-energy (2.102) in terms of $G_k(t)$ and the response function $\dot{G}_k(t)$. [After Deker and Haake (1975b).]

ISBN 0-8053-6610-5, 0-8053-6611-3 (pbk.)

tion on the right-hand side of (2.108) is obtained by taking all pairings of the four variables. In this approximation, the right-hand side is zero, which is consistent with the fact that the lowest-order term in the self-energy is proportional to g_0^2.

Deker and Haake (1975b) discuss how the expansion of $\Sigma_k(t)$ can be reorganized so as to incorporate exact static correlation functions, much as we have attempted in (2.107). Their procedure also leads to the disappearance of an explicit dependence of the self-energy on the coupling constant, and they demonstrate the value of their reorganization by comparing various approximations with exact numerical data for a simple wavevector independent version of (2.83).

APPENDIX 2A

We shall denote the local temperature at the position **r** in an incompressible liquid by $T(\mathbf{r})$. It is assumed that the temperature differences between different parts of the liquid are sufficiently small for the motion of the atoms to be unaffected by the inhomogeneities in local temperature. At the same time, the heating resulting from internal friction is taken to be a negligible effect by comparison with the variations in the local temperature. Under these circumstances, called forced convection, $T(\mathbf{r})$ satisfies the ordinary equation for heat conduction in a moving liquid,

$$(\partial_t + \mathbf{v}\cdot\boldsymbol{\nabla})T = \kappa_0 \nabla^2 T \tag{2A.1}$$

where κ_0 is the coefficient of thermal diffusivity. The occurrence of the velocity $\mathbf{v}$ in (2A.1) is a purely kinematic effect. Notice that the form of (2A.1) is identical with the diffusion equation that describes the variation in concentration of some admixture in the liquid that does not influence the motions of the atoms, i.e. if the concentration field is $c(\mathbf{r}, t)$ it satisfies an equation like (2A.1), with $T \to c$, and κ_0 is interpreted as the coefficient of molecular diffusion.

The spatial Fourier components of the local temperature $T_{\mathbf{k}}$ are defined by the transformation

$$T(\mathbf{r}) = \frac{\Omega}{(2\pi)^3} \int d\mathbf{k}\; T_{\mathbf{k}} \exp(-i\mathbf{k}\cdot\mathbf{r}), \tag{2A.2}$$

and since $T(\mathbf{r})$ is real, $T_{\mathbf{k}}^* = T_{-\mathbf{k}}$. Using the definition (2.24) for the Fourier components of the velocity field, we obtain from (2A.1) the following equation-of-motion for $T_{\mathbf{k}}$,

$$(\partial_t + \kappa_0 k^2)T_{\mathbf{k}} = i(\chi/\Omega)^{1/2} \sum_{\mathbf{pq}} \delta_{\mathbf{k},\mathbf{p}+\mathbf{q}} \sum_{\alpha} p^{\alpha} a_{\mathbf{q}}^{\alpha} T_{\mathbf{p}}. \tag{2A.3}$$

Defining the temperature autocorrelation function

$$E_k(t) = \langle T_{\mathbf{k}}(t) T_{\mathbf{k}}^* \rangle \tag{2A.4}$$

ISBN 0-8053-6610-5, 0-8053-6611-3 (pbk.)

we obtain from (2A.3) the equation

$$(\partial_t + \kappa_0 k^2)E_k(t) = i(\chi/\Omega)^{1/2} \sum_{\mathbf{pq}} \delta_{\mathbf{k,p+q}} \sum_{\alpha} p^\alpha \langle a_{\mathbf{q}}^\alpha T_{\mathbf{p}} T_{\mathbf{k}}^*(-t)\rangle$$

$$= -\frac{\chi}{\Omega} \sum_{\mathbf{pq}} \sum_{\mathbf{p'q'}} \delta_{\mathbf{k,p+q}} \delta_{\mathbf{k,p'+q'}} \sum_{\alpha\beta} p^\alpha p'^\beta$$

$$\times \int_0^t d\bar{t} \exp\{-\kappa_0 k^2(|t| - |\bar{t}|)\}$$

$$\times \langle a_{\mathbf{q}}^\alpha(\bar{t}) a_{\mathbf{q'}}^{\beta *} T_{\mathbf{p}}(\bar{t}) T_{\mathbf{p'}}^* \rangle. \quad (2A.5)$$

The second equality in (2A.5) follows by using the solution for $T_{\mathbf{k}}^*(-\bar{t})$ obtained from (2A.3). Equation (2A.5) is an exact equation for $E_k(t)$, and it is the analogue of (2.33) for the velocity autocorrelation function.

The right-hand side of (2A.5) is approximated by replacing the correlation function by the product of $\langle a_{\mathbf{q}}^\alpha(\bar{t}) a_{\mathbf{q'}}^{\beta *}\rangle$ and $\langle T_{\mathbf{p}}(\bar{t}) T_{\mathbf{p'}}^*\rangle$, i.e. we neglect the correlation between the temperature and velocity variables. At the same time we replace the exponential in the integrand by $E_k(t - \bar{t})/\langle T^2\rangle$, where $\langle T^2\rangle = E_k(0)$ is taken to be a constant independent of the wavevector. After making these approximations, and using (2.35) for the velocity autocorrelation function we obtain, finally,

$$\tilde{E}_k(s) = \int_0^\infty dt \exp(-st) E_k(t)$$

$$= \langle T^2\rangle / \{s + \kappa_0 k^2 + \tilde{\Sigma}_k^T(s)\} \quad (2A.6)$$

where the collisional self-energy

$$\tilde{\Sigma}_k^T(s) = \frac{k^2\chi}{\Omega\langle T^2\rangle} \sum_{\mathbf{p}} \{1 - (\hat{\mathbf{k}}\cdot\hat{\mathbf{p}})^2\} \int_0^\infty d\bar{t} \exp(-s\bar{t}) E_{k-p}(\bar{t}) g_p(\bar{t}) \quad (2A.7)$$

and $\hat{\mathbf{k}} = \mathbf{k}/k$, $\hat{\mathbf{p}} = \mathbf{p}/p$. The factor k^2 in (2A.7) occurs because the local temperature is a conserved variable.

Consider now the calculation of the self-energy correct to first order in χ. This is accomplished by replacing $E_{k-p}(\bar{t})$ and $g_p(\bar{t})$ in (2A.7) by their noninteracting values, namely,

$$E_{k-p}(\bar{t}) \simeq \langle T^2\rangle \exp\{-\kappa_0 |\mathbf{k} - \mathbf{p}|^2 \bar{t}\}, \quad (2A.8)$$

and $g_p(\bar{t}) \simeq \exp(-\nu_0 p^2 \bar{t})$, with the result

$$\tilde{\Sigma}_k^T(s) \simeq \frac{k^2\chi}{(2\pi)^3} \int d\mathbf{p}\, \{1 - (\hat{\mathbf{k}}\cdot\hat{\mathbf{p}})^2\}(s + \kappa_0|\mathbf{k} - \mathbf{p}|^2 + \nu_0 p^2)^{-1}, \quad (2A.9)$$

which is independent of $\langle T^2\rangle$. In going from (2A.7) to (2A.9) we have replaced the sum over $\mathbf{p}$ by an integral.

ISBN 0-8053-6610-5, 0-8053-6611-3 (pbk.)

The hydrodynamic limit of (2A.9) is obtained by setting $\mathbf{k} = 0$ in the integrand and taking the limit $s \to 0$. On setting $\mathbf{k} = 0$ in the integrand the integration over $\hat{\mathbf{p}}$ can be performed and it contributes a factor $8\pi/3$. If the maximum wavevector in the integral is Λ, the hydrodynamic limit of (2A.9) is readily shown to be

$$\tilde{\Sigma}_k^T(s) \simeq \frac{k^2\chi}{3\pi^2(\kappa_0 + \nu_0)} \left\{ \Lambda - \frac{\pi}{2} (s/(\kappa_0 + \nu_0))^{1/2} \right\} . \qquad (2A.10)$$

From this result we deduce that the first-order correction to the thermal diffusivity due to the mixing term in Eq. (2A.1) is independent of the mean-square fluctuation in the temperature $\langle T^2 \rangle$, and its temperature dependence is therefore determined largely by χ. The long-time tail induced in the thermal diffusivity by the velocity field is of exactly the same form as the corresponding term in the kinematic viscosity. The amplitude of this term in the thermal diffusivity depends on κ_0 and ν_0 since it is induced by the velocity field.

ISBN 0-8053-6610-5, 0-8053-6611-3 (pbk.)

CHAPTER III

GENERALIZED LANGEVIN EQUATION

In this chapter we illustrate the use of a generalized Langevin equation due to Mori (1965a) which is not restricted to the description of dynamic properties at large distances and long times. In fact, Mori's generalized Langevin equation is an exact, linear equation for the variables of interest. The essential feature of the equation is a separation of the force into components that have widely separated time scales. In this respect, it is identical with the simple Langevin equation discussed in Sec. 2.1 where the frictional force, which is proportional to the self-diffusion constant, has a time scale that is much longer than the time scale for the random force which is determined by the mean time between atomic collisions. The range of validity of a Langevin equation is limited by the time scale of the random force; for Brownian motion the simple Langevin equation is valid for the description of processes that occur on a time scale which is much larger than the mean time between atomic collisions.

In the study of the dynamic properties of condensed matter the variables of interest are often wavevector dependent variables that are conserved in the limit of zero wavevector. Hence, for small wavevectors, or long wavelengths, the variables are almost conserved. For the sake of brevity, we call a variable which is conserved in the limit of long wavelengths simply a conserved variable. In Sec. 2.2 we discussed the nature of the autocorrelation function of a conserved variable in the limit of long wavelengths and times. We expect macroscopic behavior of the autocorrelation function of a conserved variable for $k \to 0$ and $t \to \infty$, by which we mean that the autocorrelation function decays exponentially at long times with a decay time that tends to infinity as $k \to 0$.

Linear hydrodynamic equations-of-motion can be derived from the generalized Langevin equation by starting with the basic conserved variables and taking the limit of large distances and long times. In view of this,

Stephen Lovesey, Condensed Matter Physics: Dynamic Correlations

ISBN 0-8053-6610-5, 0-8053-6611-3 (pbk.)

it is very appealing to describe short-distance and short-time processes in terms of generalized hydrodynamic equations with generalized transport coefficients which reduce to the appropriate macroscopic transport coefficients in the limit of large distances and long times.

The form of the exact generalized Langevin equation, derived in Sec. 3.1, for a variable A is

$$\dot{A} \equiv \partial_t A(t) = i\Omega \cdot A(t) - \int_0^t d\bar{t}\, M(t-\bar{t}) \cdot A(\bar{t}) + f(t). \tag{3.1}$$

Explicit expressions for the frequency Ω, memory function $M(t)$ and random force $f(t)$ are given in later sections. An important property of $f(t)$ is that it is uncorrelated with A at all times. Moreover, the memory function $M(t)$ is proportional to the spectrum of the random force, and therefore, with an appropriate choice for $A(t)$, its time scale is short compared to that of $A(t)$. If $M(t)$ decays in a time τ_m, and the time of interest is much longer than τ_m, then $M(t) = 2L\delta(t)$ to a good approximation and the generalized Langevin equation (3.1) reduces to the same form as the simple Langevin equation for Brownian motion. Moreover, we shall find that for $t \gg \tau_m$, $\langle A(t)A^* \rangle \sim \exp((i\Omega - L)t)$ so that if Re $L \ll \tau_m^{-1}$ we have the correct description of the slow relaxation of the variable A.

The generalized Langevin equation derived in Sec. 3.1 is applied to four problems in Secs. 3.2 and 3.4–3.6. In the first problem we calculate effects, on the vibrational spectrum of a continuum model of a solid, due to weak anharmonic corrections to the Hamiltonian. This type of problem is customarily tackled with an equation-of-motion or Green's function method. However, it can equally well be tackled with the formalism of Mori's generalized Langevin equation, as we shall demonstrate in Sec. 3.2. The problem of weak anharmonic corrections is also a relatively simple problem that nicely illustrates some important features of the various quantities that enter the generalized Langevin equation. The remaining three problems treated in this chapter are formulated so that correct hydrodynamic behavior is obtained in the long-wavelength and long-time limit. This goal is achieved by including the basic conserved variables amongst the variables A in the generalized Langevin equation.

In the hydrodynamic limit we obtain results for the transport coefficients in the form of integrals over all time of the zero wavevector limit of correlation functions formed from the nonconserved parts of the fluxes, i.e. the components of the fluxes in the conservation equations that are orthogonal to the conserved variables. It is shown that these results are identical with the standard Kubo formulae for transport coefficients. An alternative approach to the study of transport phenomena is based on the construction of appropriate Boltzmann equations; see, for example, Ziman (1964), Chap. 7. This approach and the generalized Langevin equation studied here lead

ISBN 0-8053-6610-5, 0-8053-6611-3 (pbk.)

to identical results for the transport coefficients, as expected. A comparative study of the two approaches is given by Schotte (1978).

3.1 GENERALIZED LANGEVIN EQUATION

The proof of the equation-of-motion (3.1) with $f(t)$ and $M(t)$ having the described properties hinges on the use of a projection operator P. In order to define P we must define the function of principal interest, i.e. an auto-correlation function or a relaxation function. Because we would like the equation-of-motion to apply equally well for classical and quantum variables we introduce a general form for the function as a scalar product, and the notation,

$$(A(t), A^+).$$

For a classical system,

$$(A(t), A^+) = \beta\langle\{A(t) - \langle A\rangle\}\{A^+ - \langle A^+\rangle\}\rangle \tag{3.2}$$

and, in the quantal case,

$$(A(t), A^+) = R(t) \tag{3.3}$$

where $R(t)$ is the relaxation function defined by (1.55). The scalar product satisfies,

$$(A, B) = (B^+, A^+)^*, \tag{3.4}$$

$$(A, A^+) \geq 0, \tag{3.5}$$

and

$$(A, B) = (B, A). \tag{3.6}$$

If we are dealing with spatial Fourier components of local Hermitian variables, denoted by A_{k} and B_{k}, say, then translational invariance requires,

$$(A_{\mathrm{p}}(t), B_{\mathrm{q}}^+) = \delta_{\mathrm{p,q}}(A_{\mathrm{p}}(t), B_{\mathrm{p}}^+). \tag{3.7}$$

Inversion symmetry requires,

$$(A_{\mathrm{k}}(t), B_{\mathrm{k}}^+) = (A_{-\mathrm{k}}(t), B_{-\mathrm{k}}^+) \tag{3.8}$$

and since $A_{\mathrm{k}}^+ = A_{-\mathrm{k}}$ the scalar product $(A_{\mathrm{k}}(t), A_{\mathrm{k}}^+)$ is purely real.

The symmetry property of the scalar product under time reversal is concisely expressed in terms of the time reversal signatures of the dynamic variables. Let A and B be either even or odd with respect to time reversal, which means that they are either purely real or imaginary, respectively. The signature of A, ϵ_A, is $+1$ or -1 according to whether A is even or odd under time reversal. In the most general case we must consider the possibility that the system is subject to a magnetic field $\mathbf{H}$. The time reversal property of

ISBN 0-8053-6610-5, 0-8053-6611-3 (pbk.)

the Hamiltonian is then $\mathcal{H}(\mathbf{H}) \to \mathcal{H}(-\mathbf{H})$, and

$$(A(t), B)_{\mathrm{H}} = \epsilon_A \epsilon_B (A(-t), B)_{-\mathrm{H}}. \tag{3.9}$$

If $B = A$, and A is Hermitian then, for $\mathbf{H} = 0$,

$$(A(t), A) = (A(-t), A) = (A, A(t)). \tag{3.10}$$

If A is a single variable, A and $\dot{A}$ have opposite time reversal signature and, for $\mathbf{H} = 0$,

$$(\dot{A}, A) = 0. \tag{3.11}$$

Notice that for a quantum system the last result follows directly from the general relation (1.51b),

$$(\dot{A}, B) = -i\langle [A, B] \rangle. \tag{3.12}$$

The projection operator P is defined as the projection of B, say, onto the A axis,

$$PB = (B, A^+)\cdot(A, A^+)^{-1}\cdot A. \tag{3.13}$$

P is linear and Hermitian,

$$(PA, B^+) = (A, (PB)^+) \tag{3.14}$$

and idempotent

$$P^2 = P. \tag{3.15}$$

The dot notation used in (3.13) is introduced to facilitate the extension of our derivations to the case when A represents several variables. In this instance, A is a column matrix, and A^+ is its Hermitian row matrix. (B, A^+) is a matrix, and the dots in (3.13) denote matrix multiplication. $(A(t), A^+)^+$ is the Hermitian conjugate of a square matrix.

We now address the problem of deriving the generalized Langevin equation (3.1). The variable of interest is separated into two parts

$$A(t) = PA(t) + (1 - P)A(t) = F(t)\cdot A + A'(t) \tag{3.16}$$

where, from (3.13),

$$F(t) = (A(t), A^+)\cdot(A, A^+)^{-1} \tag{3.17}$$

and the second equality in (3.16) defines the component $A'(t)$ which, by definition, is orthogonal to $A(t)$ for all times. Equation (3.16) is consistent with (3.1) if the Laplace transform of $F(t)$ satisfies

$$\tilde{F}(s) = \int_0^\infty dt \exp(-st)F(t) = \{s - i\Omega + \tilde{M}(s)\}^{-1} \tag{3.18}$$

ISBN 0-8053-6610-5, 0-8053-6611-3 (pbk.)

and

$$A'(t) = \int_0^t d\bar{t}\, F(\bar{t})\cdot f(t-\bar{t}). \tag{3.19}$$

From (3.19) we deduce that $f(t)$ is orthogonal to A for all times, $t \geq 0$, as required.

We now verify (3.18) and obtain explicit expressions for $f(t)$ and $M(t)$. To this end we form an equation-of-motion for $A'(t)$ in (3.17). At this stage it is convenient to introduce the Liouville operator $\mathscr{L}$ which satisfies

$$\partial_t A(t) = \dot{A}(t) = i\mathscr{L}A(t)$$

and (3.20)

$$A(t) = \exp(it\mathscr{L})A.$$

For a classical system $\mathscr{L}$ is expressed in terms of a Poisson bracket [Résibois and de Leener (1977)] and for a quantal system $\mathscr{L}A = [\mathscr{H}, A]$. The Liouville operator is Hermitian,

$$(\mathscr{L}A, B^+) = (A, (\mathscr{L}B)^+) \tag{3.21}$$

from which it follows that,

$$(\dot{A}, B^+) = -(A, \dot{B}^+) \tag{3.22}$$

and

$$(A(t), B^+) = (A, B^+(-t)). \tag{3.23}$$

From the definition of $A'(t)$, Eqs. (3.16) and (3.20),

$$\begin{aligned}\partial_t A'(t) &= i(1-P)\mathscr{L}\{F(t)\cdot A + A'(t)\}\\ &= i(1-P)\mathscr{L}A'(t) + F(t)\cdot i(1-P)\mathscr{L}A.\end{aligned} \tag{3.24}$$

Noting that $A'(0) = 0$, we find from this last equation the result (3.19) with

$$f = i(1-P)\mathscr{L}A = (1-P)\dot{A},$$

and

$$f(t) = \exp\{it(1-P)\mathscr{L}\}\, f \tag{3.25}$$

and

$$(f(t), A^+) = 0. \tag{3.26}$$

An interesting feature of the random force is that its temporal evolution is given by the operator $\exp\{it(1-P)\mathscr{L}\}$ which differs from the evolution operator for the dynamical variables by the presence of the projection operator. We shall discuss the significance of this difference after completing the derivation of Eq. (3.18) for $F(t)$.

ISBN 0-8053-6610-5, 0-8053-6611-3 (pbk.)

From the definition of $F(t)$, Eq. (3.17), and using (3.16) and (3.23), we find

$$\begin{aligned}\dot{F}(t) &= (\dot{A}, A^+(-t))\cdot(A, A^+)^{-1}\\ &= (\dot{A}, \{PA(-t) + (1 - P)A(-t)\}^+)\cdot(A, A^+)^{-1}\\ &= (\dot{A}, \{(A(-t), A^+)\cdot(A, A^+)^{-1}\cdot A\}^+)\cdot(A, A^+)^{-1}\\ &\quad + (\dot{A}, \{A'(-t)\}^+)\cdot(A, A^+)^{-1}.\end{aligned}$$

In the first term,

$$F^+(-t) = (A, A^+)^{-1}\cdot F(t)\cdot(A, A^+) \tag{3.27}$$

and in the second term we use (3.19) and the fact that $P\dot{A}$ is orthogonal to $A'(t)$ for all times; then

$$\begin{aligned}\dot{F}(t) &= (\dot{A}, A^+)\cdot(A, A^+)^{-1}\cdot F(t)\\ &\quad + \int_0^{-t} d\bar{t}\, ((1 - P)\dot{A}, f^+(-t - \bar{t}))\cdot F^+(\bar{t})\cdot(A, A^+)^{-1}\\ &= i\Omega\cdot F(t) - \int_0^{t} d\bar{t}\, M(t - \bar{t})\cdot F(\bar{t}).\end{aligned} \tag{3.28}$$

Here we have defined the frequency matrix

$$i\Omega = \dot{F}(0) = (\dot{A}, A^+)\cdot(A, A^+)^{-1} \tag{3.29}$$

and the memory function matrix,

$$\begin{aligned}M(t) &= ((1 - P)\dot{A}, f^+(-t))\cdot(A, A^+)^{-1}\\ &= (f(t), f^+)\cdot(A, A^+)^{-1}\end{aligned} \tag{3.30}$$

where the second equality in (3.30) follows from the definition of the random force (3.25) and the fact that the operator $(1 - P)$ is Hermitian with respect to operators which are orthogonal to A, e.g. $(1 - P)\dot{A}$. Noting that $F(0) = 1$, the Laplace transform of $F(t)$ from Eq. (3.28) is exactly (3.18), and the derivation of the equation-of-motion for $F(t)$ has given explicit expressions for the frequency and memory function matrices, Eqs. (3.29) and (3.30).

It is instructive to look at an alternative derivation of the generalized Langevin equation that proceeds directly from the Laplace transform

$$\tilde{F}(s) = (\{s - i\mathscr{L}\}^{-1}A, A^+)\cdot(A, A^+)^{-1}. \tag{3.31}$$

Using the operator identity,

$$(A + B)^{-1} = A^{-1} - A^{-1}\cdot B\cdot(A + B)^{-1}, \tag{3.32}$$

ISBN 0-8053-6610-5, 0-8053-6611-3 (pbk.)

we find

$$(s - i\mathscr{L})^{-1} = (s - i\mathscr{L}(1 - P) - i\mathscr{L}P)^{-1}$$
$$= (s - i\mathscr{L}(1 - P))^{-1}$$
$$- (s - i\mathscr{L}(1 - P))^{-1}\cdot(-i\mathscr{L}P)\cdot(s - i\mathscr{L})^{-1}. \tag{3.33}$$

Because $(1 - P)A = 0$, it follows that

$$\{s - i\mathscr{L}(1 - P)\}^{-1}\cdot A = A/s. \tag{3.34}$$

The relation equivalent to (3.17) is

$$P\cdot(s - i\mathscr{L})^{-1}\cdot A = \tilde{F}(s)\cdot A. \tag{3.35}$$

Combining (3.33)—(3.35) together with (3.31) we obtain the following equation for $\tilde{F}(s)$,

$$s\tilde{F}(s) = 1 + \tilde{F}(s)\cdot(s\{s - i\mathscr{L}(1 - P)\}^{-1}i\mathscr{L}A, A^+)\cdot(A, A^+)^{-1}$$
$$= 1 + \tilde{F}(s)\cdot(\{s - i\mathscr{L}(1 - P) + i\mathscr{L}(1 - P)\}$$
$$\times \{s - i\mathscr{L}(1 - P)\}^{-1}\dot{A}, A^+)\cdot(A, A^+)^{-1}$$
$$= 1 + \tilde{F}(s)\cdot i\Omega + \tilde{F}(s)\cdot(i\mathscr{L}(1 - P)$$
$$\times \{s - i\mathscr{L}(1 - P)\}^{-1}\dot{A}, A^+)\cdot(A, A^+)^{-1}$$
$$= 1 + \tilde{F}(s)\cdot i\Omega - \tilde{F}(s)\cdot((1 - P)\{s - i\mathscr{L}(1 - P)\}^{-1}\dot{A},$$
$$\{(1 - P)\dot{A}\}^+)\cdot(A, A^+)^{-1}$$
$$= 1 + \tilde{F}(s)\cdot i\Omega - \tilde{F}(s)\cdot(\{s - i(1 - P)\mathscr{L}\}^{-1}(1 - P)\dot{A},$$
$$\{(1 - P)\dot{A}\}^+)\cdot(A, A^+)^{-1} \tag{3.36}$$

where in reaching the last equality we have used, (3.22), $(1 - P)^2 = (1 - P)$, the Hermitian property of $(1 - P)$, and

$$(1 - P)\{s - i\mathscr{L}(1 - P)\}^{-1} = \{s - i(1 - P)\mathscr{L}\}^{-1}(1 - P).$$

From this identity it is straightforward to verify (3.26) in terms of Laplace transformed quantities since,

$$\int_0^\infty dt \exp(-st)(f(t), A^+)$$
$$= (\{s - i(1 - P)\mathscr{L}\}^{-1}(1 - P)\dot{A}, A^+)$$
$$= ((1 - P)\{s - i\mathscr{L}(1 - P)\}^{-1}\dot{A}, A^+)$$
$$= (\{s - i\mathscr{L}(1 - P)\}^{-1}\dot{A}, \{(1 - P)A\}^+) = 0. \tag{3.36a}$$

With the definition (3.25) of the random force, the Laplace transform

ISBN 0-8053-6610-5, 0-8053-6611-3 (pbk.)

of the memory matrix (3.30) is

$$\tilde{M}(s) = \int_0^\infty dt \exp(-st)M(t)$$

$$= (\{s - i(1 - P)\mathscr{L}\}^{-1}(1 - P)\dot{A}, \{(1 - P)\dot{A}\}^+)\cdot(A, A^+)^{-1}. \quad (3.37)$$

The left-hand side is precisely the quantity which appears in the last equality in (3.36), and, in consequence, we recover the result (3.18) for $\tilde{F}(s)$.

Let us now investigate the consequences of the special temporal evolution operator associated with the random force. Following Mori (1965a), we examine the variable

$$I(t) = \exp(it\mathscr{L})\, f = \dot{A}(t) - i\Omega\cdot A(t) \quad (3.38)$$

which differs from $f(t)$, Eq. (3.19), only in the form of the time development operator.

We shall now establish a useful relation between $(I(t), I^+)$ and $(f(t), f^+)$. It is possible to proceed either directly from (3.1) or in terms of Laplace transformed quantities, as in (3.36). We choose the latter route, and the interested reader might like to verify the final result, (3.40), starting from (3.1).

From (3.38) and (3.33),

$$\int_0^\infty dt \exp(-st)(I(t), I^+) = (\{s - i\mathscr{L}\}^{-1}f, f^+)$$

$$= (\{s - i\mathscr{L}(1 - P)\}^{-1}f, f^+)$$

$$+ (\{s - i\mathscr{L}(1 - P)\}^{-1}i\mathscr{L}P\{s - i\mathscr{L}\}^{-1}f, f^+)$$

$$= \tilde{M}(s)\cdot(A, A^+) + (\{s - i\mathscr{L}\}^{-1}f, A^+)$$

$$\times (A, A^+)^{-1}\cdot\tilde{M}(s)\cdot(A, A^+)$$

$$= \{\tilde{M}(s) - \tilde{M}(s)\cdot\tilde{F}(s)\cdot\tilde{M}(s)\}\cdot(A, A^+)$$

where in reaching the last line we have used (3.18). The desired form of the result is

$$\int_0^\infty dt \exp(-st)\{(I(t), I^+)\cdot(A, A^+)^{-1}\} = \tilde{M}(s)\cdot\{1 - \tilde{F}(s)\cdot\tilde{M}(s)\} \quad (3.39)$$

or, alternatively,

$$(I(t), I^+)\cdot(A, A^+)^{-1}$$

$$= M(t) - \int_0^t d\bar{t} \int_0^{\bar{t}} d\bar{t}'\, M(t - \bar{t})\cdot F(\bar{t} - \bar{t}')\cdot M(\bar{t}'). \quad (3.40)$$

ISBN 0-8053-6610-5, 0-8053-6611-3 (pbk.)

Let us for the moment assume that, for the times of interest,

$$M(t) = 2L\delta(t), \qquad t \gg \tau_m.$$

In this case (3.40) reduces to ($t \gg \tau_m$)

$$(I(t), I^+)\cdot(A, A^+)^{-1} = 2L\delta(t) - L^2F(t), \tag{3.41}$$

from which it is evident that the autocorrelation function of $I(t)$ on the left-hand side includes the slow processes contained in $F(t)$. The effect of the projection operator in the time development of $f(t)$ is to remove these slow processes from the evolution of the random force.

3.2 ANHARMONIC VIBRATIONS IN A CONTINUUM MODEL OF A SOLID

The reasons for examining weak anharmonic effects in a solid are really twofold. First, the problem is obviously one of great physical interest. The second reason is to demonstrate how a perturbative calculation is carried out in terms of the generalized Langevin equation. We shall go beyond the mean-field approximation discussed in Sec. 1.7 and calculate phonon damping. In doing so, we hope also to make the reader familiar with the various quantities that enter the Langevin equation by calculating them for a relatively simple model.

The present problem can be tackled equally well with the equation-of-motion methods described in Chap. II, or with the use of causal Green's functions, Chap. V. The interested reader might like to derive the results obtained here by these alternative methods.

If we denote the displacement of an atom from its equilibrium position by u, and the conjugate momentum by p, the Hamiltonian for a single atom of unit mass is taken to be of the form

$$\mathscr{H} = \tfrac{1}{2}(p^2 + \omega_0^2 u^2) - \lambda_0 u^3 - g_0 u^4. \tag{3.42}$$

Here, ω_0 is the harmonic frequency and λ_0 and g_0 are the strengths of the cubic and quartic anharmonic terms. In generalizing the classical Hamiltonian (3.42) to describe the motions of atoms in a crystal we shall continue to assume that λ_0 and g_0 are small scalar constants, independent of the positions of the atoms, which is reasonable for certain optic lattice vibrations.

The phonon dispersion in the crystal is denoted by $\omega_{\mathbf{k}} = \omega_{-\mathbf{k}}$ and the density of wavevectors $\mathbf{k}$ is $\Omega/(2\pi)^3$. As usual, it is convenient to introduce phonon creation and annihilation operators $a_{\mathbf{k}}^+$ and $a_{\mathbf{k}}$, respectively, which satisfy

$$[a_{\mathbf{p}}, a_{\mathbf{q}}^+] = \delta_{\mathbf{pq}}, \tag{3.43}$$

all other commutators being zero. It is also convenient in the ensuing

ISBN 0-8053-6610-5, 0-8053-6611-3 (pbk.)

calculation to use the displacement operator

$$Q_{\mathrm{k}} = (1/2\omega_k)^{1/2}(a_{\mathrm{k}} + a^+_{-\mathrm{k}}) \tag{3.44}$$

and the conjugate momentum operator

$$P_{\mathrm{k}} = -i(\omega_k/2)^{1/2}(a_{\mathrm{k}} - a^+_{-\mathrm{k}}). \tag{3.45}$$

Note that $Q^+_{\mathrm{k}} = Q_{-\mathrm{k}}$ and $P^+_{\mathrm{k}} = P_{-\mathrm{k}}$. From (3.43) it follows that,

$$[Q_{\mathrm{p}}, P^+_{\mathrm{q}}] = i\delta_{\mathrm{pq}}. \tag{3.46}$$

The displacement field $u(\mathbf{r})$ is given in terms of the displacement operator by the transformation

$$u(\mathbf{r}) = \sum_{\mathrm{k}} Q_{\mathrm{k}} \exp(i\mathbf{k}\cdot\mathbf{r}) = \frac{\Omega}{(2\pi)^3}\int d\mathbf{k}\, Q_{\mathrm{k}} \exp(i\mathbf{k}\cdot\mathbf{r}), \tag{3.47}$$

and this keeps $u(\mathbf{r})$ real, as can be readily verified. Our model continuum, quantal Hamiltonian for lattice vibrations is obtained from (3.42), and it takes the form* [Ziman (1969)]

$$\begin{aligned} \mathcal{H} &= \sum_{\mathrm{k}} \omega_k a^+_{\mathrm{k}} a_{\mathrm{k}} - \lambda_0 \Omega^{-1} \int d\mathbf{r}\, u^3(\mathbf{r}) - g_0 \Omega^{-1} \int d\mathbf{r}\, u^4(\mathbf{r}) \\ &= \sum_{\mathrm{k}} \omega_k a^+_{\mathrm{k}} a_{\mathrm{k}} - \lambda_0 \sum_{123} \delta_{1+2+3,0} Q_1 Q_2 Q_3 \\ &\quad - g_0 \sum_{1234} \delta_{1+2+3+4,0} Q_1 Q_2 Q_3 Q_4. \end{aligned} \tag{3.48}$$

In (3.48) we have introduced for brevity the shorthand notation for wavevectors $\mathbf{k}_1 = 1$, etc.

Using (3.48) we obtain the following equations-of-motion for Q_{k} and P_{k},

$$\dot{Q}_{\mathrm{k}} = -i[Q_{\mathrm{k}}, \mathcal{H}] = P_{\mathrm{k}} \tag{3.49}$$

and

$$\begin{aligned} \dot{P}_{\mathrm{k}} &= -\omega_k^2 Q_{\mathrm{k}} + 3\lambda_0 \sum_{23} \delta_{k,2+3} Q_2 Q_3 \\ &\quad + 4g_0 \sum_{234} \delta_{k,2+3+4} Q_2 Q_3 Q_4. \end{aligned} \tag{3.50}$$

ISBN 0-8053-6610-5, 0-8053-6611-3 (pbk.)

* For a realistic model of lattice vibrations in a crystal the coefficients of the cubic and quartic terms in the continuum model (3.48) are modified so that (a) the sum of the wavevectors equals a reciprocal lattice vector, and (b) $1 \to \mathbf{k}_1, j_1$ where j_1 labels the various phonon branches.

We now construct a generalized Langevin equation in terms of the two-component variable,

$$A = \{Q_k, P_k\} \tag{3.51}$$

where the right-hand side denotes a column matrix, and A^+ is the corresponding, Hermitian conjugate, row matrix. A matrix of relaxation functions is defined, in accord with (3.17), whose Laplace transform $\tilde{F}(s)$ satisfies (3.18). The (1,1) element of $\tilde{F}(s)$ is proportional to the Laplace transform of the relaxation function

$$(Q_k(t), Q_k^+) = \int_0^\beta d\tau \,\langle Q_k^+(-i\tau)Q_k(t)\rangle - \beta\langle Q_k^+\rangle\langle Q_k\rangle. \tag{3.52}$$

In order to calculate the frequency matrix Ω_k, (3.29), we must calculate the elements of the matrices $(\dot{A}, A^+)$ and (A, A^+). Because Q_k and P_k have opposite time reversal signatures,

$$(Q_k, P_k^+) = 0. \tag{3.53}$$

Using (3.12), (3.46) and (3.49)

$$(P_k, P_k^+) = (\dot{Q}_k, P_k^+) = -i\langle[Q_k, P_k^+]\rangle = 1 \tag{3.54}$$

and, from (3.22)

$$(\dot{P}_k, Q_k^+) = -1. \tag{3.55}$$

Writing,

$$\chi_k = (Q_k, Q_k^+) \tag{3.56}$$

we find

$$i\Omega_k = \begin{pmatrix} 0 & 1 \\ -\chi_k^{-1} & 0 \end{pmatrix}. \tag{3.57}$$

The last result is required in the calculation of the memory function matrix, (3.30) or (3.37). The random force (3.25) is, using (3.49),

$$\begin{aligned} f &= \{\dot{Q}_k, \dot{P}_k\} - i\Omega_k \cdot \{Q_k, P_k\} \\ &= \{0, \dot{P}_k + \chi_k^{-1}Q_k\}. \end{aligned} \tag{3.58}$$

We now examine the calculation of χ_k. From the definition of χ_k, Eq. (3.56) and (3.50),

$$\begin{aligned} \omega_k^2\chi_k = 1 &+ 3\lambda_0 \sum_{23} \delta_{k,2+3}(Q_2Q_3, Q_k^+) \\ &+ 4g_0 \sum_{234} \delta_{k,2+3+4}(Q_2Q_3Q_4, Q_k^+), \end{aligned}$$

which is an exact equation. Using (3.50) again, we substitute for Q_k^+ on the right-hand side. The terms in the resulting expression which are nominally

ISBN 0-8053-6610-5, 0-8053-6611-3 (pbk.)

of order $g_0\lambda_0^*$ are multiplied by a relaxation function with an odd number of Q's and, to leading order, this is proportional to λ_0, making the complete term of order $g_0|\lambda_0|^2$. In consequence, if we work to order $|\lambda_0|^2$ and g_0^2,

$$\begin{aligned}\omega_k^4\chi_k = \omega_k^2 &+ 3\lambda_0 \sum_{23} \delta_{k,2+3}(Q_2Q_3, \\ &- \dot{P}_k^+ + 3\lambda_0^* \sum_{2'3'} \delta_{k,2'+3'} Q_{2'}^+Q_{3'}^+) \\ &+ 4g_0 \sum_{234} \delta_{k,2+3+4}(Q_2Q_3Q_4, \\ &- \dot{P}_k^+ + 4g_0^* \sum_{2'3'4'} \delta_{k,2'+3'+4'} Q_{2'}^+Q_{3'}^+Q_{4'}^+).\end{aligned} \tag{3.59}$$

From (3.12) and (3.46),

$$(\dot{P}_k^+, Q_2Q_3) = -i\langle[P_k^+, Q_2Q_3]\rangle = 0 \tag{3.60}$$

and

$$\begin{aligned}(\dot{P}_k^+, Q_2Q_3Q_4) &= -i\langle[P_k^+, Q_2Q_3Q_4]\rangle \\ &= -\langle\delta_{2,k}Q_3Q_4 + \delta_{3,k}Q_2Q_4 + \delta_{4,k}Q_2Q_3\rangle.\end{aligned} \tag{3.61}$$

Because (3.61) appears in χ_k multiplied by λ_0 we shall evaluate the static correlation functions on the right-hand side of (3.61) using the harmonic part of the Hamiltonian. With the notation

$$\langle a_k^+ a_k\rangle = \{\exp(\beta\omega_k) - 1\}^{-1} = n_k \tag{3.62}$$

we find

$$\begin{aligned}(\dot{P}_k, Q_2Q_3Q_4) = -\frac{1}{2}\Big\{&\delta_{2,k}\delta_{3,-4}\left(\frac{2n_3+1}{\omega_3}\right) \\ &+ \delta_{3,k}\delta_{2,-4}\left(\frac{2n_2+1}{\omega_2}\right) + \delta_{4,k}\delta_{2,-3}\left(\frac{2n_2+1}{\omega_2}\right)\Big\}.\end{aligned} \tag{3.63}$$

The next term we shall evaluate is the coefficient of $|\lambda_0|^2$. Because we shall later need the time-dependent version of the relaxation function involved we evaluate

$$\begin{aligned}(Q_2(t)Q_3(t), Q_{2'}^+Q_{3'}^+) &= -\beta\langle Q_2Q_3\rangle\langle Q_{2'}^+Q_{3'}^+\rangle \\ &+ \int_0^\beta d\tau\,\langle Q_{2'}^+(-i\tau)Q_{3'}^+(-i\tau)Q_2(t)Q_3(t)\rangle.\end{aligned} \tag{3.64}$$

The first term on the right-hand side follows from the definition of the relaxation function (1.55). For a harmonic system

$$Q_2(t) = \left(\frac{1}{2\omega_2}\right)^{1/2}\{a_2\exp(-i\omega_2 t) + a_{-2}^+\exp(i\omega_2 t)\} \tag{3.65}$$

ISBN 0-8053-6610-5, 0-8053-6611-3 (pbk.)

and,*

$$\langle Q_{2'}^{+}(-i\tau)Q_{3'}^{+}(-i\tau)Q_2(t)Q_3(t)\rangle$$
$$= \langle Q_{2'}^{+}Q_{3'}^{+}\rangle \langle Q_2Q_3\rangle + (\delta_{22'}\delta_{33'} + \delta_{2'3}\delta_{23'})$$
$$\times \langle Q_2^{+}(-i\tau)Q_2(t)\rangle \langle Q_3^{+}(-i\tau)Q_3(t)\rangle. \tag{3.66}$$

From these two results we find,

$$(Q_2(t)Q_3(t),\ Q_2^{z}\,Q_3^{z})$$
$$= (\delta_{22}\,\delta_{33} + \delta_{2\,3}\delta_{23})\,\frac{1}{4\omega_2\omega_3}$$
$$\times \left\{\left(\frac{1 + n_2 + n_3}{\omega_2 + \omega_3}\right) [\exp\{-it(\omega_2 + \omega_3)\} + \exp\{it(\omega_2 + \omega_3)\}]\right.$$
$$\left. + \left(\frac{n_3 - n_2}{\omega_2 - \omega_3}\right) [\exp\{it(\omega_3 - \omega_2)\} + \exp\{it(\omega_2 - \omega_3)\}]\right\}. \tag{3.67}$$

In obtaining the form (3.67) we have used relations like

$$n_2n_3\{\exp[\beta(\omega_2 + \omega_3)] - 1\} = 1 + n_2 + n_3.$$

When we set $t = 0$ in (3.67) we obtain the desired approximation for the coefficient of the $|\lambda_0|^2$ in χ_k.

The coefficient of g_0^2 in the expression (3.59) for χ_k involves a sixth-order relaxation function, which means that it represents three phonon processes. We shall neglect this term and work to order g_0 and $|\lambda_0|^2$, i.e. we retain the first-order contributions from the cubic and quartic terms in the Hamiltonian. With these approximations

$$\chi_k^{-1} = \omega_k^2 - 6g_0 \sum_2 (2n_2 + 1)/\omega_2$$
$$- 9|\lambda_0|^2 \sum_{23} \delta_{k,2+3}\left(\frac{1}{\omega_2\omega_3}\right)$$
$$\times \left\{\left(\frac{1 + n_2 + n_3}{\omega_2 + \omega_3}\right) + \left(\frac{n_3 - n_2}{\omega_2 - \omega_3}\right)\right\}. \tag{3.68}$$

The second two terms are proportional to temperature in the limit of high temperatures. The term of order g_0 in (3.68) is of precisely the same form as the mean-field correction to the frequency derived in Sec. 1.7.

---

* If the variables A_m have a Gaussian distribution then $\langle A_1A_2\cdots A_{2n}\rangle = \Sigma\langle A_{l_1}A_{l_2}\rangle\langle A_{l_3}A_{l_4}\rangle\cdots\langle A_{l_{2n-1}}A_{l_{2n}}\rangle$, where the sum extends over all possible divisions of the $2n$ indices, 1, 2, 3, . . . , $2n$, into pairs. There are 1, 3, . . . , $(2n - 1)$ terms in the sum.

ISBN 0-8053-6610-5, 0-8053-6611-3 (pbk.)

We now use the result (3.68) in conjunction with (3.50) to calculate the one element of the random force in (3.58). If we again limit ourselves to two phonon processes, then

$$\dot{P}_{\mathrm{k}} + \chi_k^{-1} Q_{\mathrm{k}} \simeq 3\lambda_0 \sum_{23} \delta_{k,2+3} Q_2 Q_3 . \tag{3.69}$$

In view of the fact that the random force is proportional to the strength of the cubic anharmonic term, and therefore a small quantity, we neglect the projection operator in the time evolution of the random force. This means that we approximate the memory function matrix $M(t)$ by $(I(t), I^+)(A, A^+)^{-1}$ where I is defined in (3.38). The nonzero element of $M(t)$ is then equal to,

$$9|\lambda_0|^2 \sum_{23} \delta_{k,2+3} \sum_{2'3'} \delta_{k,2'+3'} (Q_2(t) Q_3(t),\ Q_{2'}^+ Q_{3'}^+), \tag{3.70}$$

and the relaxation function appearing in this expression is given by (3.67). If we denote the Laplace transform of (3.70) by $\tilde{M}_{22}(s)$, the memory matrix is

$$\tilde{M}(s) = \begin{pmatrix} 0 & 0 \\ 0 & \tilde{M}_{22}(s) \end{pmatrix} . \tag{3.71}$$

The matrix of relaxation functions is calculated from (3.18) using (3.57) and (3.71) for the frequency and memory function matrices. From the resulting expression we obtain the result

$$\int_0^\infty dt \exp(-st)(Q_{\mathrm{k}}(t), Q_{\mathrm{k}}^+) = \frac{\chi_k(s + \tilde{M}_{22}(s))}{s^2 + \chi_k^{-1} + s\tilde{M}_{22}(s)} \tag{3.72}$$

and the corresponding causal Green's function has a Fourier transform, cf. Eqs. (1.64) and (1.66),

$$2\pi\langle\langle Q_{\mathrm{k}};\ Q_{\mathrm{k}}^+\rangle\rangle = -\{s^2 + \chi_k^{-1} + s\tilde{M}_{22}(s)\}^{-1}, \tag{3.73}$$

with $s = i\omega + \eta$. The denominator in (3.72) and (3.73) is, to order g_0 and λ_0^2,

$$\begin{aligned} s^2 + \chi_k^{-1} + s\tilde{M}_{22}(s) \simeq{}& s^2 + \omega_k^2 - 6g_0 \sum_2 (2n_2 + 1)/\omega_2 \\ & - i9|\lambda_0|^2 \sum_{23} \delta_{k,2+3} \left(\frac{1}{2\omega_2\omega_3}\right) \\ & \times \{(1 + n_2 + n_3)[(s + i\omega_2 + i\omega_3)^{-1} \\ & - (s - i\omega_2 - i\omega_3)^{-1}] \\ & + 2(n_3 - n_2)(s + i\omega_2 - i\omega_3)^{-1}\}. \end{aligned} \tag{3.74}$$

The frequency dependent terms are evaluated to first-order with s set equal to $i\omega_k + \eta$. The quartic interaction, treated in the mean-field approximation,

ISBN 0-8053-6610-5, 0-8053-6611-3 (pbk.)

and the real part of the first-order contribution from the cubic interaction together result in a temperature and wavevector dependent shift to the bare, harmonic phonon frequency. We also find, at this level of approximation, a temperature and wavevector dependent damping of the phonons coming from two-phonon collision processes. As was remarked earlier, the contribution of order g_0^2 involves the collision of three phonons, and it contributes to both the shift and damping of the phonons. It is easy to see that, in the limit if high temperatures the damping term in (3.74) is proportional to the frequency and temperature. This dependence of the damping on temperature merely reflects the fact that the number of thermally excited phonons participating in the collision process increases with temperature. We are led to conclude that phonon damping is largest for high-frequency phonons and high temperatures.

The structure of (3.74) is exactly the same as that obtained in the study of small anharmonic contributions to the phonon Green's function by diagrammatic perturbation theory; see, for example, Barron and Klein (1974).

The Hamiltonian for a realistic model of lattice vibrations is more complicated than our continuum model, but the additional complexities do not modify the structure of the phonon self-energy. Let us focus on the calculation of the phonon lifetime which comes from the last term in (3.74). For a model of a crystal the most significant modifications to (3.74) are (a) the sum of the wavevectors is equal to $\mathbf{k}$ plus a reciprocal lattice vector, (b) the spatial Fourier transform of the third-order derivative of the potential [represented in (3.74) by a constant, λ_0] introduces an additional wavevector dependence and (c) $1 \rightarrow \mathbf{k}_1, j_1$ where j_1 labels the various phonon branches (longitudinal and transverse acoustic, for example, for one atom per unit cell and wavevectors along high symmetry directions). In the limit of high temperatures the phonon damping is predicted to vary linearly with the temperature, it is proportional to the square of the Grüneisen parameter, and the variation with wavevector is proportional to the square of the phonon dispersion for simple crystals.

We have assumed throughout our discussion that the anharmonic terms are small perturbations. When the anharmonicity is sufficiently strong additional phenomena occur such as second sound which is a propagating collective mode in the energy density. The present formalism can be used to describe strongly anharmonic crystals by employing a different type of approximation scheme to the one used here that is similar in many respects to the high-frequency approximation introduced in our subsequent discussion of a paramagnet [see, for example, Schneider and Stoll (1978)].

3.3 CONTINUED FRACTION EXPANSION

In equation-of-motion methods for calculating correlation functions, one attempts to calculate the nonlinear terms in the equations-of-motion for A,

ISBN 0-8053-6610-5, 0-8053-6611-3 (pbk.)

say, by constructing the equation-of-motion for the required nonlinear terms. This latter equation will, of course, be even "more nonlinear" than the equation-of-motion for A, and an approximation must be made at some stage to terminate the hierarchy of equations. The purpose of constructing the higher-order equations-of-motion is to find a reasonable solution for the original equation for A.

Nonlinear effects in the Langevin equation are contained in the memory function. An explicit example is given in the preceding section where it is shown that in the calculation of the relaxation function for nonlinear lattice vibrations the random force is proportional to the cubic interaction. It is possible to construct an equation for the memory function which has the same form as the generalized Langevin equation for $F(t)$. Hence, we have an infinite chain of equations for a series of memory functions, the lowest order of which determines the correlation function of interest. The chain of equations for $\tilde{F}(s)$ has the form of a continued fraction* [Mori (1965b)].

To see how this comes about, we return to the problem discussed in the preceding section, where the two-component variable A has the special form

$$A = \{Q_k, \dot{Q}_k\}. \tag{3.75}$$

A consequence of the fact that the second variable is the derivative of the first is that the random force has only one component,

$$f = \{0, \ddot{Q}_k + \chi_k^{-1}Q_k\}, \tag{3.76}$$

and from this result it follows that the memory function matrix has one element

$$\tilde{M}_{22}(s) = ((s - i(1 - P)\mathscr{L})(\ddot{Q}_k + \chi_k^{-1}Q_k), (\ddot{Q}_k + \chi_k^{-1}Q_k)^+)(\dot{Q}_k, \dot{Q}_k^+)^{-1}. \tag{3.77}$$

In (3.77) we have, for the sake of completeness, explicitly included the factor $(\dot{Q}_k, \dot{Q}_k^+) = 1$.

The result (3.72) for the Laplace transform of $(Q_k(t), Q_k^+)$ can be written

$$\int_0^\infty dt \exp(-st)(Q_k(t), Q_k^+) = \chi_k(s + \tilde{M}(s))^{-1} \tag{3.78}$$

where the new memory function (not a matrix) is,

$$M(s) = \chi_k^{-1}(\dot{Q}_k, \dot{Q}_k^+)(s + \tilde{M}_{22}(s))^{-1}. \tag{3.79}$$

It is easy to show from (3.78) that for small times,

$$(Q_k(t), Q_k^+) = \chi_k\{1 - \tfrac{1}{2}t^2 M(t = 0) + \cdots\} \tag{3.80}$$

ISBN 0-8053-6610-5, 0-8053-6611-3 (pbk.)

* An alternative continued fraction scheme is described by Karasudani et al. (1979).

so that

$$M(0) = -(\ddot{Q}_{\mathrm{K}}, Q_{\mathrm{K}}^{+})\chi_{k}^{-1} = (\dot{Q}_{\mathrm{K}}, \dot{Q}_{\mathrm{K}}^{+})\chi_{k}^{-1}. \tag{3.81}$$

In view of this result, it follows from (3.79) that $M(t)$ satisfies an equation

$$\partial_t M(t) = -\int_0^t d\bar{t}\, M_{22}(t - \bar{t})M(\bar{t}). \tag{3.82}$$

We conclude from (3.82) that $M_{22}(t)$ is the memory function of the memory function $M(t)$.

It is no doubt evident that if we use a variable $A = \{Q_{\mathrm{K}}, \dot{Q}_{\mathrm{K}}, \ddot{Q}_{\mathrm{K}}, \ldots\}$ then we generate an infinite chain of equations of which (3.78) and (3.79) are the first two members; to be precise we obtain a continued fraction expansion for the memory function $\tilde{M}(s)$. Each additional term in the continued fraction expansion for $\tilde{M}(s)$ corresponds to the inclusion of an additional time derivative of Q_{K} in the column matrix variable A.

We now summarize the properties of the continued fraction expansion. For this purpose, let A denote a single variable, and

$$F(t) = (A(t), A^{+})(A, A^{+})^{-1}. \tag{3.83}$$

The continued fraction expansion for the Laplace transform of $F(t)$ is then

$$\tilde{F}(s) = (s + \tilde{M}^{(1)}(s))^{-1} \tag{3.84}$$

where

$$\tilde{M}^{(n)}(s) = \delta_n/(s + \tilde{M}^{(n+1)}(s)). \tag{3.85}$$

The coefficient δ_n in (3.85) is the initial value of $M^{(n)}(t)$, and it can be expressed in terms of the moments of Fourier transform

$$F(\omega) = \frac{1}{2\pi}\int_{-\infty}^{\infty} dt\, \exp(-i\omega t)F(t). \tag{3.86}$$

The nth moment is defined to be

$$\langle \omega^n \rangle = \int_{-\infty}^{\infty} d\omega\, \omega^n F(\omega), \tag{3.87}$$

and since $F(t)$ is an even function of t, the nonzero moments have n equal to an even integer. Moreover, the short-time expansion of $F(t)$ is,

$$F(t) = 1 - t^2\langle \omega^2 \rangle/2! + t^4\langle \omega^4 \rangle/4! - \cdots. \tag{3.88}$$

With this notation the first three δ's are,

$$\begin{aligned} \delta_1 &= \langle \omega^2 \rangle, \\ \delta_2 &= (\langle \omega^4 \rangle/\langle \omega^2 \rangle) - \langle \omega^2 \rangle, \\ \delta_2\delta_3 &= (\langle \omega^6 \rangle/\langle \omega^2 \rangle) - (\langle \omega^4 \rangle/\langle \omega^2 \rangle)^2. \end{aligned} \tag{3.89}$$

ISBN 0-8053-6610-5, 0-8053-6611-3 (pbk.)

It is useful to note that, if $F(t)$ is a Gaussian

$$F(t) = \exp(-\tfrac{1}{2}\sigma^2 t^2) \tag{3.90}$$

then

$$\langle \omega^n \rangle = \frac{n!}{(n/2)!}\left(\frac{1}{2}\right)^n \sigma^n \tag{3.91}$$

and

$$\delta_n = n\sigma^2. \tag{3.92}$$

If the continued fraction expansion is terminated at the nth level by replacing $\tilde{M}^{(n+1)}(s)$ by a constant, independent of s, then the first n moments of $F(t)$ are correct.

3.4 PARAMAGNET IN AN APPLIED FIELD

A paramagnet consists of an assembly of magnetic moments (or spins) in which the average magnetic moment is zero, i.e. there is no long-range magnetic order. For simple three-dimensional (3D) systems a paramagnetic state is achieved at temperatures large compared to the strength of spin-spin interactions, and spin ordering (ferromagnetic, antiferromagnetic, etc.) occurs continuously as the temperature is decreased below a critical temperature which is proportional to the strength of the spin-spin interaction. If the spin-spin interaction is spatially isotropic 1D and 2D systems do not display long-range magnetic order, although very strong short-range order sets in at temperatures small compared to the strength of spin-spin interactions.

Because we require wavelength dependent variables that reduce to macroscopic variables in the long-wavelength limit our starting point is the theory of the thermodynamic properties of a paramagnet. It is well known, for example, that the isothermal susceptibility is related to the mean-square fluctuation in the magnetization. Couched in slightly different terms, the susceptibility is a measure of the response of a paramagnet to fluctuations in the magnetization. Similarly, the specific heat at constant magnetization is found to be a measure of temperature fluctuations. We shall assume that both the local magnetization and energy are conserved variables.

Recall that the thermodynamic state of a single component system is specified completely by two variables. One variable is chosen to be the magnetization, M, and it is convenient to choose the second variable in such a way that it is uncorrelated with M. We shall construct such a variable from the energy density, and we show that its mean-square thermodynamic fluctuation is proportional to the specific heat at constant magnetization. Altogether there are four response functions, we can regard the specific

ISBN 0-8053-6610-5, 0-8053-6611-3 (pbk.)

heats at constant magnetization and field as "thermal" response functions, and the isothermal and isentropic susceptibilities as "mechanical" response functions.

We begin our brief view of thermodynamic fluctuations in a paramagnet by defining the appropriate Gibbs distribution* [Callen (1960)]. The internal energy and magnetization in the direction of the applied field are denoted by E and M, respectively. The average value of a variable at a temperature $T = 1/k_B\beta$ is calculated from a distribution

$$p_j = Z^{-1}\exp\{\beta(hM_j - E_j)\} \tag{3.93}$$

where the index j labels values of E and M, and the partition function Z is defined by the requirement that $\Sigma\, p = 1$. The field h in p_j is the derivative of E with respect to M at constant entropy, i.e.

$$h = (\partial E/\partial M)_s \quad \text{and} \quad T = (\partial E/\partial S)_m. \tag{3.94}$$

By definition, the magnetization

$$M = \sum_j p_j M_j \tag{3.95}$$

and the isothermal susceptibility [Stanley (1971)]

$$\chi_t = (\partial M/\partial h)_T. \tag{3.96}$$

Noting that,

$$(\partial Z/\partial h)_T = \beta Z M \tag{3.97}$$

we find

$$\chi_t = \beta \sum_j p_j(M_j - M)^2 = (M, M). \tag{3.98}$$

The magnetization fluctuations within a finite volume of the paramagnet are small except in the vicinity of the phase transition below which M is finite even in zero field. The specific heat at constant field C_h is the derivative of the magnetic enthalpy W with respect to temperature at constant field. With the definition,

$$W = E - hM$$

we find,

$$TC_h = T(\partial W/\partial T)_h = (W, W). \tag{3.99}$$

The specific heat at constant magnetization C_m is,

$$C_m = (\partial E/\partial T)_m. \tag{3.100}$$

The natural variable here is, in the sense that M and W are the natural

* Volume and pressure effects in the paramagnet are assumed to be negligible.

ISBN 0-8053-6610-5, 0-8053-6611-3 (pbk.)

variables for χ_t and C_h, respectively,

$$\mathscr{E} = E - (E, M)M/\chi_t \tag{3.101}$$

and

$$TC_m = (\mathscr{E}, \mathscr{E}). \tag{3.102}$$

Notice that $\mathscr{E}$ is the component of the internal energy which is orthogonal to the magnetization, i.e. $(\mathscr{E}, M) = 0$. The proof of (3.101) and (3.102) is obtained from the identity,

$$\left(\frac{\partial E}{\partial \beta}\right)_m = \left(\frac{\partial E}{\partial \beta}\right)_h - \left(\frac{\partial E}{\partial h}\right)_T \left(\frac{\partial M}{\partial \beta}\right)_h \chi_t^{-1} \tag{3.103}$$

and the results,

$$\beta(\partial M/\partial \beta)_h = h\chi_t - (E, M), \tag{3.104}$$

$$(\partial E/\partial h)_T = (E, M) \tag{3.105}$$

and,

$$\beta(\partial E/\partial \beta)_h = h(E, M) - (E, E). \tag{3.106}$$

Finally, the isentropic susceptibility

$$\chi_s = (\partial M/\partial h)_s \tag{3.107}$$

is obtained from the identity,

$$C_h(\chi_t - \chi_s) = T(\partial M/\partial T)_h^2 . \tag{3.108}$$

It is easily verified that,

$$\gamma = C_h/C_m = \chi_t/\chi_s = 1 + [h\chi_t - (E, M)]^2/(\mathscr{E}, \mathscr{E})\chi_t \geq 1. \tag{3.109}$$

In zero field $(E, M) = 0$, since E and M have opposite time reversal signatures. In consequence, the ratio γ is unity in zero field. For a finite field the deviation of γ from unity is, clearly, a measure of the coupling induced between the energy and magnetization.

In view of the preceding results for the thermodynamic fluctuations in a paramagnet, we define wavevector dependent variables M_{k},

$$W_{\mathrm{k}} = E_{\mathrm{k}} - hM_{\mathrm{k}} \tag{3.110}$$

and

$$\mathscr{E}_{\mathrm{k}} = E_{\mathrm{k}} - (E_{\mathrm{k}}, M_{\mathrm{k}}^+)M_{\mathrm{k}}\chi_t(k)^{-1} \tag{3.111}$$

and the functions

$$\chi_t(k) = (M_{\mathrm{k}}, M_{\mathrm{k}}^+), \tag{3.112}$$

$$TC_h(k) = (W_{\mathrm{k}}, W_{\mathrm{k}}^+) \tag{3.113}$$

ISBN 0-8053-6610-5, 0-8053-6611-3 (pbk.)

and

$$TC_m(k) = (\mathscr{E}_{\mathrm{k}}, \mathscr{E}^{+}_{\mathrm{k}}). \tag{3.114}$$

In the limit of long wavelengths, (3.112), (3.113) and (3.114) reduce, respectively, to the isothermal susceptibility, and specific heats at constant field and magnetization. Equation (3.114) can be written in a more appealing form in terms of a Fourier component of a local temperature variable

$$T_{\mathrm{k}} = \mathscr{E}_{\mathrm{k}}/C_m(k) \tag{3.115}$$

since then,

$$C_m(k) = T(T_{\mathrm{k}}, T^{+}_{\mathrm{k}})^{-1}, \tag{3.116}$$

which is of the same form as the corresponding result for specific heat of a liquid at constant volume.*

To complete the description of the static fluctuations, we introduce a local thermodynamic field,

$$H_{\mathrm{k}} = M_{\mathrm{k}}\chi_t^{-1}(k) + [h - (E_{\mathrm{k}}, M^{+}_{\mathrm{k}})\chi_t^{-1}(k)]T_{\mathrm{k}}/T \tag{3.117}$$

and

$$\chi_s(k) = (H_{\mathrm{k}}, H^{+}_{\mathrm{k}})^{-1}, \tag{3.118}$$

reduces to the thermodynamic result in the long-wavelength limit.

The choice of (3.117) as the local field is substantiated by considering the thermodynamic fluctuation in h in terms of M and T. Using standard thermodynamic results, the fluctuation in the field, Δh, in a finite volume of the paramagnet is

$$\begin{aligned} \Delta h &= (\partial h/\partial M)_T \Delta M + (\partial h/\partial T)_m \Delta T \\ &= (\Delta M - \alpha \Delta T)\chi_t^{-1} \end{aligned} \tag{3.119}$$

where

$$T\alpha = T(\partial M/\partial T)_h = (E, M) - h\chi_t. \tag{3.120}$$

Hence,

$$\Delta h = \Delta M \chi_t^{-1} + \{h - (E, M)\chi_t^{-1}\}\Delta T/T, \tag{3.121}$$

and if we identify Δh, ΔM and ΔT with H_{k}, M_{k} and T_{k}, respectively, then (3.117) and (3.121) have exactly the same structure.

A similar discussion for the fluctuation in the entropy permits us to

* If we isolate a finite volume in a liquid, or gas, then the volume will experience temperature fluctuations ΔT where $\langle (\Delta T)^2 \rangle = T^2/\Omega C_V$, and C_V is the specific heat at constant volume. We identify T_{k} in (3.116) with ΔT, and refer to T_{k} as a Fourier component of a local temperature variable.

ISBN 0-8053-6610-5, 0-8053-6611-3 (pbk.)

identify W_k as a local entropy. We find for the fluctuation in entropy

$$T\Delta S = C_m \Delta T + (\alpha T/\chi_t)\Delta M. \tag{3.122}$$

This result, together with (3.120), suggests that we define a local entropy variable

$$S_k = W_k/T, \tag{3.123}$$

in terms of which

$$C_h(k) = T(S_k, S_k^+). \tag{3.124}$$

The variables M_k, $\mathscr{E}_k$, and H_k, W_k are orthogonal pairs of variables, in the sense that

$$(M_k, \mathscr{E}_k^+) = 0 \tag{3.125}$$

and

$$(H_k, W_k^+) = 0. \tag{3.126}$$

The variables M_k, H_k and T_k, S_k (or $\mathscr{E}_k$, W_k) can justifiably be regarded as the basic wavelength dependent "mechanical" and "thermal" variables for a paramagnet.

To the best of our knowledge, none of the response functions can be calculated exactly for arbitrary wavevectors for a three-dimensional model paramagnet. The best one can do for a Heisenberg paramagnet is to expand the Gibbs distribution in powers of β, and such high temperature series expansion quickly become very complicated after the first couple of terms.

A model which does yield to analysis is the one-dimensional, classical Heisenberg magnet [Mattis (1965), Steiner *et al.* (1976)]. Because several quasi-one-dimensional compounds have been studied extensively by a variety of techniques, a one-dimensional magnet is not quite such an academic subject to study as it might appear at first sight. As is well known, a one-dimensional magnet with an isotropic interaction between the spins does not order at a finite temperature, i.e. it is a paramagnet for all temperatures.

One of the most interesting observations made with 1D magnets is the existence of a collective oscillation of the magnetization at finite wavevectors, which shows up in the spectral density of magnetizaion fluctuations as a peak at a nonzero frequency; an example is shown in Fig. (3.1). The essential physics behind this phenomenon is that, although there is no long-range magnetic order, there is very strong short-range order in a one-dimensional magnet at temperatures small compared to the strength of the interaction between spins. Hence, if the magnet is probed, at a sufficiently short wavelength, it is possible to excite a collective oscillation of the magnetization, albeit short-lived, much as it occurs in an ordered magnet (cf. Sec. 5.4). The substantial short-range order does not influence significantly the long-wavelength properties of a 1D magnet, and for $\mathbf{k} \to 0$ the

ISBN 0-8053-6610-5, 0-8053-6611-3 (pbk.)

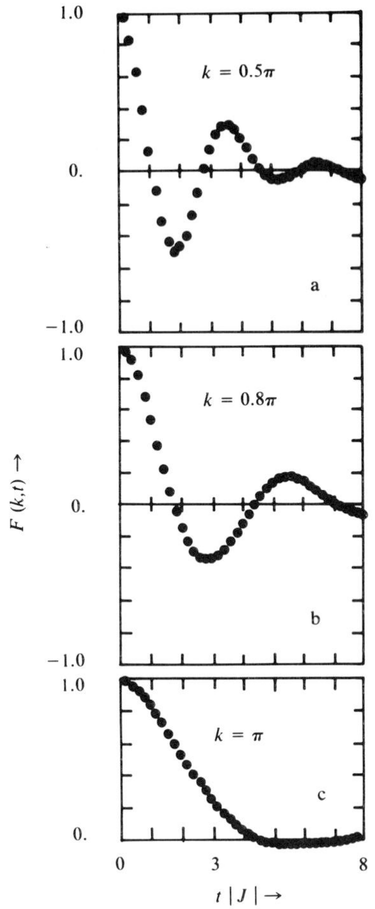

Fig. (3.1) The figure shows computer simulation data for an antiferromagnetically coupled linear chain of classical spins at a temperature $T^* = 0.3|J|$ where J is the nearest neighbor exchange coupling, and zero applied field. [After Loveluck and Windsor (1978).] In Figs. (a)—(c), the quantity $F(k, t) = (M_k(t), M_k^+)/(M_k, M_k^+)$ is shown as a function of time for three values of k, and Figs. (d) and (e) show the Fourier transform, $F(k, \omega)$, for $k = 0.3\pi$ and 0.55π. The wavevectors are measured in units of the inverse of the spacing between the spins.

magnetization undergoes a diffusive motion [Forster (1975)]. When a magnetic field is applied and the temperature is lowered a ferromagnetically coupled 1D magnet will order, and, in consequence, a collective oscillation of the magnetization (or spin wave) can be supported at all wavevectors.

It is worthwhile to pause at this stage in our discussion of the dynamic properties of paramagnets to look briefly at the actual form of the static response functions for a 1D ferromagnetically coupled Heisenberg magnet.

ISBN 0-8053-6610-5, 0-8053-6611-3 (pbk.)

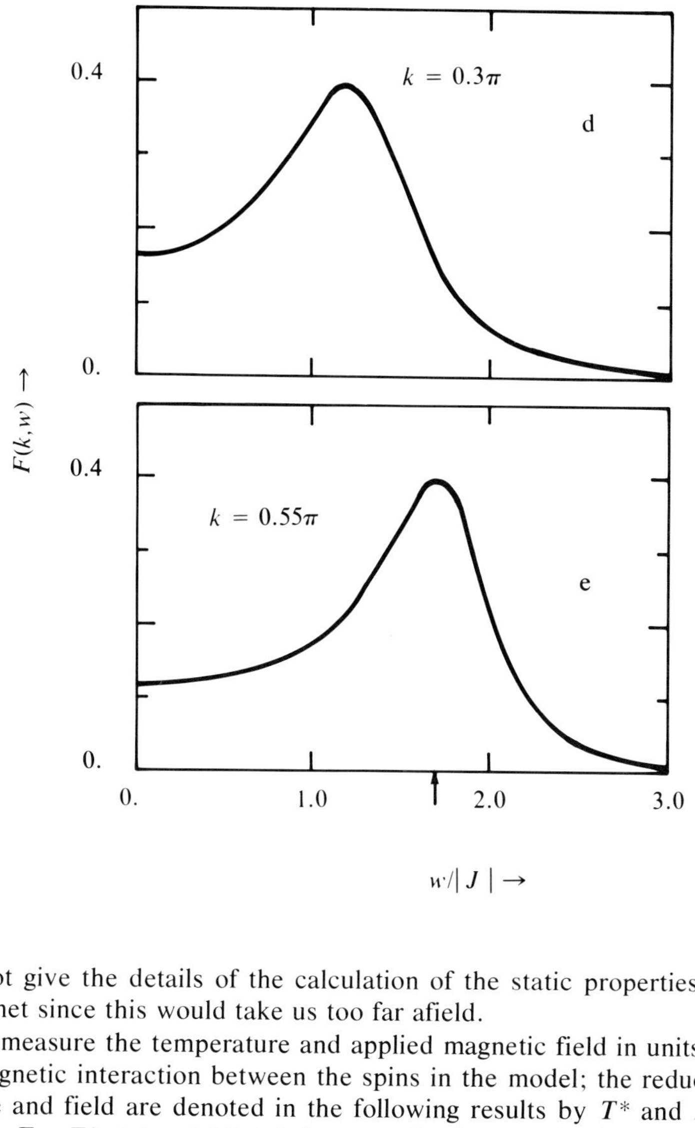

We shall not give the details of the calculation of the static properties of such a magnet since this would take us too far afield.

Let us measure the temperature and applied magnetic field in units of the ferromagnetic interaction between the spins in the model; the reduced temperature and field are denoted in the following results by T^* and h^*, respectively. For $T^* \ll 1$ and $h^* \neq 0$ the normalized magnetization is found to be [Lovesey and Loveluck (1976)]

$$M = 1 - T^*/\kappa \tag{3.127}$$

where

$$\kappa^2 = h^*(h^* + 4). \tag{3.128}$$

From (3.127) we see that the magnetization saturates at zero temperature,

ISBN 0-8053-6610-5, 0-8053-6611-3 (pbk.)

as expected. Also from (3.127),

$$\chi_t(0) = (\partial M/\partial h^*)_T = T^*(h^* + 2)/\kappa^3, \tag{3.129}$$

and this result is consistent with the result for $\chi_t(k)$ calculated from its definition as the fluctuation in M_{k}, namely,

$$\chi_t(k) = (T^*/\kappa)\left\{\frac{(h^* + 2)}{\kappa^2 + 2(1 - \cos k)}\right\}, \tag{3.130}$$

where the wavevector is measured in units of the inverse of the spacing between the spins. From (3.130) we can identify κ as the inverse of the correlation length, since $\chi_t(k)$ has the Ornstein-Zernike form for small k. The susceptibility achieves its maximum as a function of k when $\cos k = 1$, and this simply reflects the ferromagnetic nature of the coupling between the spins, since for an antiferromagnetic coupling the maximum of $\chi_t(k)$ would be achieved at $k = \pi$. In zero field, a separate calculation shows that the result which corresponds to (3.129) is*

$$\chi_t(0) = 2/3T^{*2}, \qquad T^* \ll 1. \tag{3.131}$$

Comparing this with (3.129) we see that a modest field reduces $\chi_t(0)$ very dramatically. Finally, we give the results for the specific heats ($T^* \ll 1$ and $h^* \neq 0$)

$$C_m(k) = \tfrac{1}{2}k_B\{1 - \kappa/(h^* + 2)\}(1 + \cos k) \tag{3.132}$$

and,

$$C_h(k) - C_m(k) = k_B\{h^*(\kappa - h^* - 2)(h^* + 3 + \cos k)\}^2/\kappa. \tag{3.133}$$

Notice that both specific heats are independent of the temperature, and that the ratio $C_h(0)/C_m(0)$ is very different from unity for modest values of h^*. We conclude that a significant coupling is induced between the magnetization and the energy even for modest values of the external field.

We return now to the calculation of the function $(M_{\mathrm{k}}(t), M_{\mathrm{k}}^+)$. Because, as we have seen, the magnetization and energy are coupled in the presence of a magnetic field we treat M_{k} and $\mathscr{E}_{\mathrm{k}}$ on an equal footing, and include them both in the column matrix of initial variables. Moreover, M_{k} and $\mathscr{E}_{\mathrm{k}}$ are the two basic conserved variables and must therefore be treated equally if we are to obtain a correct description of the long-wavelength properties of the paramagnet. The collective oscillation of the magnetization which can occur at short wavelengths is made evident in the formulation if we include $\dot{M}_{\mathrm{k}}$ in the initial variables, in addition to M_{k} and $\mathscr{E}_{\mathrm{k}}$. We therefore

* Because a one-dimensional magnet orders at absolute zero we expect the zero field isothermal susceptibility to diverge as $T \to 0$ since fluctuations in the magnetization become macroscopic near the phase transition.

ISBN 0-8053-6610-5, 0-8053-6611-3 (pbk.)

take,

$$A = \{M_{\mathrm{k}}, M_{\mathrm{k}}, \mathscr{E}_{\mathrm{k}}\}. \tag{3.134}$$

Before constructing the generalized Langevin equation it is instructive to examine the properties of the continuity equations for the conserved variables in (3.134). Denoting the fluxes associated with M_{k}, E_{k} and $\dot{M}_{\mathrm{k}}$ by $\mathbf{J}_{\mathrm{k}}$, $\mathbf{q}_{\mathrm{k}}$ and $\boldsymbol{\sigma}_{\mathrm{k}}$, respectively,

$$\dot{M}_{\mathrm{k}} + i\mathbf{k}\cdot\mathbf{J}_{\mathrm{k}} = 0, \tag{3.135}$$

$$\dot{E}_{\mathrm{k}} + i\mathbf{k}\cdot\mathbf{q}_{\mathrm{k}} = 0 \tag{3.136}$$

and

$$\ddot{M}_{\mathrm{k}} + i\mathbf{k}\cdot\boldsymbol{\sigma}_{\mathrm{k}} = 0. \tag{3.137}$$

We now separate the fluxes into components parallel and orthogonal to A, Eq. (3.134). For example,

$$\begin{aligned} \ddot{M}_{\mathrm{k}} &= -i\mathbf{k}\cdot\hat{\boldsymbol{\sigma}}_{\mathrm{k}} + (\ddot{M}_{\mathrm{k}}, M_{\mathrm{k}}^{+})\frac{M_{\mathrm{k}}}{\chi_t(k)} + (\ddot{M}_{\mathrm{k}}, \mathscr{E}_{\mathrm{k}}^{+})\frac{\mathscr{E}_{\mathrm{k}}}{TC_m(k)} \\ &= -i\mathbf{k}\cdot\hat{\boldsymbol{\sigma}}_{\mathrm{k}} - H'_{\mathrm{k}}(\dot{M}_{\mathrm{k}}, \dot{M}_{\mathrm{k}}^{+}) \end{aligned} \tag{3.138}$$

where the field

$$H'_{\mathrm{k}} = \frac{M_{\mathrm{k}}}{\chi_t(k)} + \frac{(\dot{M}_{\mathrm{k}}, \dot{\mathscr{E}}_{\mathrm{k}}^{+})\mathscr{E}_{\mathrm{k}}}{TC_m(k)(\dot{M}_{\mathrm{k}}, \dot{M}_{\mathrm{k}})}, \tag{3.139}$$

and $\hat{\boldsymbol{\sigma}}_{\mathrm{k}}$, by definition, is orthogonal to A. The nonconserved part of the flux $\hat{\boldsymbol{\sigma}}_{\mathrm{k}}$ contains the dissipative elements of the flux $\boldsymbol{\sigma}_{\mathrm{k}}$, and it is proportional to an element of the random force in the Langevin equation for A.*

If we neglect $\hat{\boldsymbol{\sigma}}_{\mathrm{k}}$ in (1.138) and switch off the external field, so that the second term in H'_{k} is zero, then $M_{\mathrm{k}}(t)$ varies harmonically in time with a frequency,

$$\omega_0(k) = \{(\dot{M}_{\mathrm{k}}, \dot{M}_{\mathrm{k}}^{+})\chi_t^{-1}(k)\}^{1/2}. \tag{3.140}$$

If we denote the Hamiltonian by $\mathscr{H}$, then

$$(\dot{M}_{\mathrm{k}}, \dot{M}_{\mathrm{k}}^{+}) = \langle[[M_{\mathrm{k}}, \mathscr{H}], M_{\mathrm{k}}^{+}]\rangle,$$

and from this result it follows that, for an isotropic interaction between the spins, $(\dot{M}_{\mathrm{k}}, \dot{M}_{\mathrm{k}}^{+})$ is proportional to k^2 for small wavevectors, even when the effect of an applied field is included in $\mathscr{H}$. Because $\chi_t(0)$ is a constant when $h \neq 0$, we conclude that the frequency ω_0 is proportional to k for small k and $T^* \ll 1$. For a one-dimensional magnet in a field, and low

* The random force $f = (1 - P)\dot{A}$, where P projects onto A, and so it follows from (3.135)—(3.137) that f is, apart from a factor $-i\mathbf{k}$, the nonconserved part of the fluxes.

ISBN 0-8053-6610-5, 0-8053-6611-3 (pbk.)

temperatures, the complete result is

$$\omega_0^2(k) = 2(1 - \kappa/(2 + h^*))(1 - \cos k)\{\kappa^2 + 2(1 - \cos k)\}. \tag{3.141}$$

The continuity equation for the energy (3.136) is written

$$\dot{E}_{\mathbf{k}} + \left\{\frac{(\dot{E}_{\mathbf{k}}, \dot{M}_{\mathbf{k}}^{+})\chi_t(k)}{\omega_0^2(k)}\right\} i\mathbf{k}\cdot\mathbf{J}_{\mathbf{k}} + i\mathbf{k}\cdot\hat{\mathbf{q}}_{\mathbf{k}} = 0, \tag{3.142}$$

where $\hat{\mathbf{q}}_{\mathbf{k}}$ is orthogonal to A. The content of (3.142) becomes more apparent if we introduce a variable $S'_{\mathbf{k}}$, having the dimension of entropy, which is orthogonal to $H'_{\mathbf{k}}$, namely,

$$TS'_{\mathbf{k}} = \mathscr{E}_{\mathbf{k}} - ((\mathscr{E}_{\mathbf{k}}, \dot{M}_{\mathbf{k}}^{+})/\omega_0^2(k))\chi_t(k)M_{\mathbf{k}}. \tag{3.143}$$

From (1.142) and (1.143) it is readily shown that

$$T\dot{S}'_{\mathbf{k}} + i\mathbf{k}\cdot\hat{\mathbf{q}}_{\mathbf{k}} = 0. \tag{3.144}$$

This result suggests that $\hat{\mathbf{q}}_{\mathbf{k}}$ is identified as a thermal flux, but it should be borne in mind that $S'_{\mathbf{k}}$ is not the same as the entropy variable associated with global equilibrium of the system, $S_{\mathbf{k}}$. The origin, and explanation, of this apparent inconsistency is the attempt to go beyond a description of the long-wavelength properties of a paramagnet and describe the possible collective oscillation of the magnetization at finite wavevectors by including $\dot{M}_{\mathbf{k}}$ in the initial set of variables. The long-wavelength, hydrodynamic properties of a paramagnet are described completely by the two basic conserved variables $M_{\mathbf{k}}$ and $\mathscr{E}_{\mathbf{k}}$, and the coupling between the magnetization and energy induced by the field is accounted for by renormalizing the energy, $E \to \mathscr{E}$.

A matrix of correlation functions F is defined as in (3.17) with the variable A given by (3.134). The three elements of the random force which determine the memory function matrix are

$$f_1 = 0, \qquad f_2 = -i\mathbf{k}\cdot\hat{\boldsymbol{\sigma}}_{\mathbf{k}}, \qquad f_3 = -i\mathbf{k}\cdot\hat{\mathbf{q}}_{\mathbf{k}}. \tag{3.145}$$

These results confirm our earlier comment that it is the nonconserved parts of the fluxes in the continuity equations which determine the random force. The result $f_1 = 0$ is a direct consequence of the fact that the second variable in A is the time derivative of the first variable. After some straightforward algebra we obtain

$$\tilde{F}(s)^{-1} = \begin{pmatrix} s & -1 & 0 \\ \omega_0^2(k) & s + \tilde{\theta}_2 & \tilde{\theta}_{23} + (\dot{M}_{\mathbf{k}}, \dot{\mathscr{E}}_{\mathbf{k}}^{+})/(\mathscr{E}_{\mathbf{k}}, \mathscr{E}_{\mathbf{k}}^{+}) \\ 0 & \tilde{\theta}_{32} - (\dot{M}_{\mathbf{k}}, \dot{\mathscr{E}}_{\mathbf{k}}^{+})^*/(\dot{M}_{\mathbf{k}}, \dot{M}_{\mathbf{k}}^{+}) & s + \tilde{\theta}_3 \end{pmatrix} \tag{3.146}$$

ISBN 0-8053-6610-5, 0-8053-6611-3 (pbk.)

where

$$\tilde{\theta}_2 = k^2(\dot{M}_{\mathbf{k}}, \dot{M}_{\mathbf{k}}^+)^{-1} \int_0^\infty dt \exp(-st)(\hat{\sigma}_{\mathbf{k}}(t), \hat{\sigma}_{\mathbf{k}}^+), \tag{3.147}$$

$$\tilde{\theta}_{23} = k^2(\mathscr{E}_{\mathbf{k}}, \mathscr{E}_{\mathbf{k}}^+)^{-1} \int_0^\infty dt \exp(-st)(\hat{\sigma}_{\mathbf{k}}(t), \hat{q}_{\mathbf{k}}^+), \tag{3.148}$$

$$\tilde{\theta}_{32} = k^2(\dot{M}_{\mathbf{k}}, \dot{M}_{\mathbf{k}}^+)^{-1} \int_0^\infty dt \exp(-st)(\hat{q}_{\mathbf{k}}(t), \hat{\sigma}_{\mathbf{k}}^+), \tag{3.149}$$

$$\tilde{\theta}_3 = k^2(\mathscr{E}_{\mathbf{k}}, \mathscr{E}_{\mathbf{k}}^+)^{-1} \int_0^\infty dt \exp(-st)(\hat{q}_{\mathbf{k}}(t), \hat{q}_{\mathbf{k}}^+). \tag{3.150}$$

Because these four functions are formed from the nonconserved parts of the fluxes, and the temporal evolution is governed by the modified Liouville operator which contains the projection operator, we anticipate that the integrands decay rapidly.

The function of main interest,

$$\tilde{F}_1(k, s) = (M_{\mathbf{k}}, M_{\mathbf{k}}^+)^{-1} \int_0^\infty dt \exp(-st)(M_{\mathbf{k}}(t), M_{\mathbf{k}}^+), \tag{3.151}$$

is the (1, 1) element of $\tilde{F}(s)$. From (3.146) we find

$$\tilde{F}_1(k, s) = (s + \omega_0^2(k)/\{s + \tilde{M}(k, s)\})^{-1} \tag{3.152}$$

where the memory function

$$\tilde{M}(k, s) = \tilde{\theta}_2 - \{\tilde{\theta}_{23} + (\dot{M}_{\mathbf{k}}, \dot{\mathscr{E}}_{\mathbf{k}}^+)/(\mathscr{E}_{\mathbf{k}}, \mathscr{E}_{\mathbf{k}}^+)\}$$
$$\times \{\tilde{\theta}_{32} - (\dot{M}_{\mathbf{k}}, \dot{\mathscr{E}}_{\mathbf{k}}^+)^*/(\dot{M}_{\mathbf{k}}, \dot{M}_{\mathbf{k}}^+)\}(s + \tilde{\theta}_3)^{-1}. \tag{3.153}$$

The results (3.152) and (3.153) together give an exact prescription for the Laplace transform of the magnetization autocorrelation function.

We examine first the long-wavelength properties of our result. In this limit the functions $\tilde{\theta}_{23}$ and $\tilde{\theta}_{32}$ can be neglected to a good approximation, since they are of higher order in k than the remaining two functions. We assume that the integrals in $\tilde{\theta}_2$ and $\tilde{\theta}_3$ are finite as $k \to 0$, and define transport coefficients ξ and ζ as

$$\xi = \int_0^\infty dt \lim_{k \to 0} (\hat{\sigma}_{\mathbf{k}}(t), \hat{\sigma}_{\mathbf{k}}^+), \tag{3.154}$$

and

$$T\zeta = \int_0^\infty dt \lim_{k \to 0} (\hat{q}_{\mathbf{k}}(t), \hat{q}_{\mathbf{k}}^+). \tag{3.155}$$

The identity (3.39) can be used to prove that the affect of the projection

ISBN 0-8053-6610-5, 0-8053-6611-3 (pbk.)

operator in the temporal evolution of the flux autocorrelation functions in (3.154) and (3.155) vanishes in the long-wavelength limit. In consequence, our definition of transport coefficients is identical to that given by Kubo (1957) and (1966). In order to verify this we first note that (3.39) can be written in the form,

$$\int_0^\infty dt \exp(-st)(I(t), I^+)\cdot(A, A^+)^{-1}$$

$$= (s - i\Omega)\cdot(s - i\Omega + \tilde{M})^{-1}\cdot\tilde{M}, \quad (3.39a)$$

where $I = (1 - P)\dot{A} = f$, and the time development of $I(t)$ is governed by the Liouville operator whereas $M(t)$ is proportional to the autocorrelation of $f(t) = \exp\{it(1 - P)\mathscr{L}\} f$. If A is taken to be a conserved variable $\varphi_{\mathbf{k}}$, then $f = -i\mathbf{k}\cdot\hat{\mathbf{J}}_{\mathbf{k}}$ where $\hat{\mathbf{J}}_{\mathbf{k}}$ is the nonconserved part of the flux associated with $\varphi_{\mathbf{k}}$, cf. Eq. (2.11), and $\Omega = (-\mathbf{k}\cdot\mathbf{J}_{\mathbf{k}}, \varphi_{\mathbf{k}}^+)$. Consider now the wavevector dependence of the quantities in (3.39a). Both sides of the equality are proportional to k^2, e.g. $M(t) = k^2(\hat{J}_{\mathbf{k}}(t), \hat{J}_{\mathbf{k}}^+)$, and this common factor can be canceled, whereupon both sides are proportional to the Laplace transform of the autocorrelation function of $\hat{\mathbf{J}}_{\mathbf{k}}$, and the time development of $\hat{\mathbf{J}}_{\mathbf{k}}$ on the left-hand side is governed by $\mathscr{L}$, whereas on the right-hand side it is governed by $(1 - P)\mathscr{L}$. Because Ω and $\tilde{M}$ vanish as $k \to 0$, and the two autocorrelation functions are therefore equal to one another in this limit, we conclude that the effect of the projection operator in the time development of nonconserved fluxes vanishes in the long-wavelength limit.

For small values of k, the frequency ω_0 takes the form $h \neq 0$)

$$\omega_0(k) = v_t k, \qquad k \to 0 \quad (3.156)$$

which defines the velocity v_t. Using the expressions (3.154)–(3.156) in the memory function (3.153) we obtain the result, valid for long wavelengths,

$$\tilde{M}(k, s) = (\xi/v_t^2\chi_t(0)) + \frac{k^2 v_t^2\{\Gamma(0) - 1\}}{(s + k^2\zeta/C_m(0))} \quad (3.157)$$

where, for general wavevectors,

$$\Gamma(k) = 1 + \frac{|(\dot{M}_{\mathbf{k}}, \dot{\mathscr{E}}_{\mathbf{k}}^+)|^2(M_{\mathbf{k}}, M_{\mathbf{k}}^+)}{(\mathscr{E}_{\mathbf{k}}, \mathscr{E}_{\mathbf{k}}^+)(\dot{M}_{\mathbf{k}}, \dot{M}_{\mathbf{k}}^+)^2} \geq 1. \quad (3.158)$$

The function $\Gamma(k)$ is unity for $h = 0$. For a one-dimensional, ferromagnetically coupled magnet with $T^* \ll 1$ and $h \neq 0$

$$\Gamma(k) = \tfrac{1}{2}\kappa[(h^* + 2 - \kappa)(1 + \cos k)/\{\kappa^2 + 2(1 - \cos k)\}] + 1 \quad (3.159)$$

from which we deduce that, for large fields, $\Gamma(0) = 1 + 98/h^{*2}$.

The coefficient ξ is related to the spin diffusion constant, D_s. This quantity is obtained from $\tilde{F}_1(k, s)$ as the limit (cf. the discussion in Sec.

ISBN 0-8053-6610-5, 0-8053-6611-3 (pbk.)

2.2)

$$D_s = \lim_{\omega \to 0} \omega^2 \lim_{k \to 0} k^{-2} \operatorname{Re} \tilde{F}_1(k, i\omega)$$

$$= v_t^4 \chi_t(0)/\xi. \tag{3.160}$$

The coefficient ζ enters into the energy diffusion constant, D_e. From the (3, 3) element of $\tilde{F}(s)$ we find,

$$(\mathscr{E}_{\mathbf{k}}, \mathscr{E}_{\mathbf{k}}^{+})^{-1} \int_0^{\infty} dt \exp(-st)(\mathscr{E}_{\mathbf{k}}(t), \mathscr{E}_{\mathbf{k}}^{+})$$

$$= \left(s + \tilde{\theta}_3 + \frac{s\omega_0^2(k)\{\Gamma(k) - 1\}}{\omega_0^2(k) + s\{s + \tilde{\theta}_2\}} \right)^{-1}, \tag{3.161}$$

and the energy diffusion constant,

$$D_e = \zeta / C_m(0) + D_s\{\Gamma(0) - 1\}. \tag{3.162}$$

For large fields $\Gamma(0) = 1$ to a good approximation, and D_e is then proportional to ζ.

Let us now consider what the theory predicts for small wavelengths for which the paramagnet might support a collective oscillation of the magnetization. $\Gamma(k)$ is unity at the Brillouin zone boundary since the energy and magnetization have opposite phases for this wavevector, cf. Eq. (3.159). In consequence, we can set $\Gamma(k) = 1$, to a good approximation, in the discussion of the collective mode. Because this mode is expected to occur at a relatively high frequency we can estimate the memory function matrix from its short-time behavior. It is easily shown that $(\hat{\sigma}_{\mathbf{k}}, \hat{q}_{\mathbf{k}}^{+}) = 0$, so we shall neglect $\tilde{\theta}_{23}$ and $\tilde{\theta}_{32}$. In estimating $\tilde{\theta}_2$ we shall use the short-time approximation,

$$(\hat{\sigma}_{\mathbf{k}}(t), \hat{\sigma}_{\mathbf{k}}^{+}) = (\hat{\sigma}_{\mathbf{k}}, \hat{\sigma}_{\mathbf{k}}^{+}) \exp(-t/\tau), \tag{3.163}$$

where τ is a relaxation time, as yet undetermined. With these approximations, we obtain, for short wavelengths and times,

$$\tilde{M}(k, s) \simeq (\hat{\sigma}_{\mathbf{k}}, \hat{\sigma}_{\mathbf{k}}^{+})(s + 1/\tau)^{-1} \tag{3.164}$$

and the corresponding form for $\tilde{F}_1$, Eq. (3.152), leads to the result,

$$\operatorname{Re} \tilde{F}_1(k, i\omega) = \frac{\tau\omega_0^2(\hat{\sigma}_{\mathbf{k}}, \hat{\sigma}_{\mathbf{k}}^{+})}{\{\omega\tau(\omega^2 - (\hat{\sigma}_{\mathbf{k}}, \hat{\sigma}_{\mathbf{k}}^{+}) - \omega_0^2)\}^2 + (\omega^2 - \omega_0^2)^2}. \tag{3.165}$$

From either (3.89), or directly from (3.138), we find that

$$(\hat{\sigma}_{\mathbf{k}}, \hat{\sigma}_{\mathbf{k}}) = \omega_l^2 - \omega_0^2 \tag{3.166}$$

where

$$\omega_l^2 = (\ddot{M}_{\mathbf{k}}, \ddot{M}_{\mathbf{k}}^{+})(\dot{M}_{\mathbf{k}}, \dot{M}_{\mathbf{k}}^{+})^{-1}. \tag{3.167}$$

ISBN 0-8053-6610-5, 0-8053-6611-3 (pbk.)

Because a collective mode will manifest itself in (3.165) as a peak at nonzero ω, the condition for the existence of such a mode is obtained by studying the minima of the denominator. A minimum at nonzero ω is found to exist for wavevectors and temperatures such that

$$3\omega_0^2 > \omega_l^2 = (1 - \cos k)(h^* + 2 - \kappa)\{3\kappa + (h^* + 2 - \kappa)(1 - \cos k)\}, \tag{3.168}$$

and the equality gives ω_l^2 for ferromagnetic coupling, $T^* \ll 1$, and $h^* \neq 0$. Notice that this condition is independent of the unknown relaxation time τ, which gives us additional confidence in the line of argument being pursued here. Moreover, when the condition (3.168) is well satisfied, the collective mode frequency is given by ω_l to a good approximation. We conclude that the theory does describe a collective oscillation in the magnetization density at short wavelengths, and that the dispersion of the mode is approximately ω_l; for a ferromagnetically coupled 1D magnet in a field at low temperatures, ω_l is given by (3.168), and this is in good agreement with data obtained by Loveluck and Balcar (1979) from a computer simulation with $T^* = 0.3$ and $h^* = 1.0$. In zero field, the inverse correlation length is zero (in the limit of zero temperature) and ω_l reduces to

$$\omega_l = 2(1 - \cos k)$$

which is, according to (5.110), the exact dispersion relation for spin waves in a one-dimensional magnet (measured in units of the ferromagnetic exchange).

Finally, we discuss the effect of the magnetic field on the magnetization autocorrelation function. Finite field effects appear in essentially two forms. First, static correlation functions depend on the strength of the field, as is seen, for example, in expressions (3.168) and (3.159) for $\omega_l^2(k)$ and $\Gamma(k)$, respectively. Secondly, there is a dynamic coupling of the magnetization and energy density fluctuations whose strength is proportional to $\Gamma(k) - 1$. For a ferromagnet, $\Gamma(k)$ decreases monotonically from its maximum value, $\Gamma(0)$, to unity at the zone boundary. Because the magnetization autocorrelation function is controlled by a diffusion process at long wavelengths, we conclude that the dynamic coupling of M_k and E_k can only reveal itself in the magnetization autocorrelation function at intermediate wavevectors away from the zone center and boundary. It is readily shown that for high frequencies, where damping effects in the memory function (3.153) are negligible, and intermediate wavevectors the magnetization autocorrelation function (3.152) contains a hybrid excitation arising from energy fluctuations, and this appears as in Re $\tilde{F}$ $(k, i\omega)$ as a second peak at a frequency below the spin-wave-like excitation. The hybrid excitation was observed first in computer simulation studies of classical, ferromagnetically coupled spins in a steady magnetic field by Loveluck and Balcar (1979).

ISBN 0-8053-6610-5, 0-8053-6611-3 (pbk.)

3.5 MONATOMIC LIQUID

The discussion of the dynamic properties of a monatomic liquid, at long wavelengths, in terms of a generalized Langevin equation, parallels closely the preceding discussion of a paramagnet. In view of this, the present discussion is quite brief, and we pay particular attention to addressing the similarities and differences between a paramagnet (in zero external field) and a monatomic liquid.

There are three basic conserved variables for a single-component liquid—namely, the particle density $n_{\mathbf{k}}$, the momentum density $m\mathbf{J}_{\mathbf{k}}$ and the energy density $E_{\mathbf{k}}$ [Lamb (1932), and Landau and Lifshitz (1963)]. Energy is dissipated in a liquid by internal friction and heat exchange between different parts of the liquid, and the corresponding transport coefficients are the shear viscosity coefficient, η, and the thermal conductivity, λ, respectively. The bulk viscosity η_B is concerned with the compression of a liquid, and η_B is the coefficient of the contribution to the pressure from the rate of change of volume.

In the following conservation equations for the three conserved variables, it is assumed that the atoms in the liquid interact through a pair-potential u that depends only on the separation between pairs of atoms. For example, many properties of liquids composed of rare gas atoms are described well by theories in which the pair-potential is taken to be of the so-called Lennard-Jones form,

$$u(r) \propto \{(\sigma/r)^{12} - (\sigma/r)^{6}\},$$

where σ is the separation between pairs of atoms at which the potential is zero.

It is necessary to distinguish between the components of the current density parallel and perpendicular to the wavevector, i.e. the longitudinal, J^L, and transverse, J^T, components. Because the transverse components of the current density are independent of the other variables they may be treated separately.

The microscopic particle density is

$$n(\mathbf{r}) = \sum_j \delta(\mathbf{r} - \mathbf{R}_j) \tag{3.169}$$

where $\mathbf{R}_j$ is the position of the jth atom, and therefore the $\mathbf{k}$th Fourier component of the fluctuation in the particle density is, for $\mathbf{k} \neq 0$,

$$n_{\mathbf{k}} = \int d\mathbf{r}\, \exp(-i\mathbf{k}\cdot\mathbf{r})n(\mathbf{r}) = \sum_j \exp(-i\mathbf{k}\cdot\mathbf{R}_j). \tag{3.170}$$

The conservation equation for $n_{\mathbf{k}}$ follows immediately from (3.170),

$$\partial_t n_{\mathbf{k}} + ikJ^L_{\mathbf{k}} = 0. \tag{3.171}$$

ISBN 0-8053-6610-5, 0-8053-6611-3 (pbk.)

Here, the current density is,

$$J_{\mathbf{k}}^{L,T} = \sum_j v^{L,T} \exp(-i\mathbf{k}\cdot\mathbf{R}_j), \tag{3.172}$$

where $\mathbf{v}$ is the particle velocity. The current density J^L has the special property of being both a flux and a conserved variable. The flux associated with the current density is the stress tensor $\sigma^{L,T}$, and [Hansen and McDonald (1976)]

$$m\partial_t J_{\mathbf{k}}^{L,T} + ik\sigma_{\mathbf{k}}^{L,T} = 0, \tag{3.173}$$

and

$$\sigma_{\mathbf{k}}^{\alpha\beta} = \sum_j \left\{ m v_j^\alpha v_j^\beta - \frac{1}{2}\sum_{i\neq j}\left(\frac{r_{ij}^\alpha r_{ij}^\beta}{r_{ij}^2}\right) P_{\mathbf{k}}(r_{ij})\right\} \exp(-i\mathbf{k}\cdot\mathbf{R}_j) \tag{3.174}$$

with

$$P_{\mathbf{k}}(R) = R(\partial u/\partial R)\{1 - \exp(-i\mathbf{k}\cdot\mathbf{R})\}/(i\mathbf{k}\cdot\mathbf{R}) \tag{3.175}$$

and

$$\mathbf{r}_{ij} = \mathbf{R}_i - \mathbf{R}_j.$$

In (3.173), $\sigma_{\mathbf{k}}^L$ is the diagonal component of (3.174), $\sigma_{\mathbf{k}}^{zz}$, while $\sigma_{\mathbf{k}}^T$ represents an off-diagonal element $\sigma_{\mathbf{k}}^{zx}$, say. Finally, the energy density,

$$E_{\mathbf{k}} = \sum_j \{\tfrac{1}{2}mv_j^2 + \tfrac{1}{2}\sum_{i\neq j} u(r_{ij})\} \exp(-i\mathbf{k}\cdot\mathbf{R}_j) \tag{3.175}$$

and the conservation equation

$$\partial_t E_{\mathbf{k}} + i\mathbf{k}\cdot\mathbf{q}_{\mathbf{k}} = 0 \tag{3.176}$$

defines the energy flux $\mathbf{q}_k$.

Before we go any further in our discussion, we record some standard results for the static values of autocorrelation functions of $n_{\mathbf{k}}$ and $\mathbf{J}_{\mathbf{k}}$. Our ultimate aim is to study the dynamic properties of fluctuations in the number density, and related quantities, and so it is the longitudinal components of the fluxes that enter most of the subsequent results. We shall now write $J_{\mathbf{k}}$ for $J_{\mathbf{k}}^L$ to simplify notation. Because we are concerned with a classical liquid the scalar product of two variables (3.2) is proportional to the correlation function formed with the fluctuations in the variables.

The first result of interest relates the static structure factor $S(k)$ to the mean square fluctuation in $n_{\mathbf{k}}$,

$$S(k) = \langle |n_{\mathbf{k}}|^2\rangle/N \tag{3.177}$$

where N is total number of atoms. $S(0)$ is related to the isothermal compressibility $1/B_T$ by the identity

$$S(0) = n/\beta B_T = \langle(\Delta N)^2\rangle/N, \tag{3.178}$$

ISBN 0-8053-6610-5, 0-8053-6611-3 (pbk.)

where n is the mean particle density and $\beta = 1/k_B T$, as usual. In comparing the static properties of a classical liquid to those of a paramagnet, the quantity $\beta S(k)$ is the analogue of the wavevector dependent isothermal susceptibility.

In calculating $\langle J_{\mathbf{k}} J_{\mathbf{k}}^{+} \rangle$ the key result used is that, for a classical liquid, there is no correlation between the velocities of different particles, i.e. $\langle v_i v_j \rangle = 0$ for $i \neq j$, and for a Maxwell distribution $\langle v^2 \rangle = 1/m\beta$. Using these results,

$$\langle |J_{\mathbf{k}}|^2 \rangle = N/m\beta. \tag{3.179}$$

The quantity that corresponds to $\langle |J_{\mathbf{k}}|^2 \rangle$ in a paramagnet is $(\dot{M}_{\mathbf{k}}, \dot{M}_{\mathbf{k}}^{+})$ and, unlike $\langle |J_{\mathbf{k}}|^2 \rangle$, this depends on the wavevector being proportional to k^2 for small wavevectors.

The four basic wavelength dependent variables for a monatomic liquid can be chosen to be the number density $n_{\mathbf{k}}$, pressure $P_{\mathbf{k}}$, temperature $T_{\mathbf{k}}$ and entropy $S_{\mathbf{k}}$; $n_{\mathbf{k}}$ and $P_{\mathbf{k}}$ may be regarded as "mechanical" variables, and $T_{\mathbf{k}}$ and $S_{\mathbf{k}}$ "thermal" variables. The pairs $T_{\mathbf{k}}$, $n_{\mathbf{k}}$ and $S_{\mathbf{k}}$, $P_{\mathbf{k}}$ are uncorrelated variables, and the mean-square fluctuations of $T_{\mathbf{k}}$, and $S_{\mathbf{k}}$ define wavevector dependent specific heats at constant volume and pressure, respectively, which coincide with thermodynamic results in the limit of long wavelengths. Explicit forms for the variables, and some of their important properties, are summarized in Table 3.1. Detailed discussions of the variables and their properties are given by Schofield (1975) and Copley and Lovesey (1975).

The transport coefficients η, η_B and λ can be expressed in terms of integrals of the autocorrelation functions of the nonconserved, or fluctuating, components of the fluxes $\sigma_{\mathbf{k}}^{T}$, $\sigma_{\mathbf{k}}^{L}$ and $q_{\mathbf{k}}$. A very detailed account of these relations is given by Ernst and Dorfman (1975), and the results are,

$$\eta = (\beta/\Omega) \int_0^\infty dt \lim_{k \to 0} \langle \sigma_{\mathbf{k}}^{T}(t) \sigma_{\mathbf{k}}^{T+} \rangle, \tag{3.180}$$

$$\eta_B + \tfrac{4}{3}\eta = (\beta/\Omega) \int_0^\infty dt \lim_{k \to 0} \langle \hat{\sigma}_{\mathbf{k}}^{L}(t) \hat{\sigma}_{\mathbf{k}}^{L+} \rangle, \tag{3.181}$$

and

$$\lambda T = (\beta/\Omega) \int_0^\infty dt \lim_{k \to 0} \langle \hat{\mathbf{q}}_{\mathbf{k}}(t) \cdot \hat{\mathbf{q}}_{\mathbf{k}}^{+} \rangle. \tag{3.182}$$

With these preliminaries completed, we can turn now to the study of the dynamic properties of a monatomic liquid. We shall examine only the longitudinal components of the variables since these include the fluctuations in the number density which is a quantity of prime interest in light scattering and acoustic attenuation.

ISBN 0-8053-6610-5, 0-8053-6611-3 (pbk.)

Table 3.1
Thermodynamic Variables for a Monatomic Liquid

Temperature

$$T_{\mathbf{k}} = \left(E_{\mathbf{k}} - \frac{\langle E_{\mathbf{k}} n_{\mathbf{k}}^{+} \rangle}{\langle n_{\mathbf{k}} n_{\mathbf{k}}^{+} \rangle} n_{\mathbf{k}} \right) (\Omega C_v(k))^{-1}$$

with

$$C_v(k) = \left(\langle E_{\mathbf{k}} E_{\mathbf{k}}^{+} \rangle - \frac{|\langle E_{\mathbf{k}} n_{\mathbf{k}}^{+} \rangle|^2}{\langle n_{\mathbf{k}} n_{\mathbf{k}}^{+} \rangle} \right) (\Omega k_B T^2)^{-1},$$

$$\langle T_{\mathbf{k}} T_{\mathbf{k}}^{+} \rangle = k_B T^2 (\Omega C_v(k))^{-1}, \qquad \langle T_{\mathbf{k}} n_{\mathbf{k}}^{+} \rangle = 0.$$

Pressure

$$P_{\mathbf{k}} = \sigma_{\mathbf{k}}^{L} - \hat{\sigma}_{\mathbf{k}}^{L} = \frac{\langle \sigma_{\mathbf{k}}^{L} n_{\mathbf{k}}^{+} \rangle}{\langle n_{\mathbf{k}} n_{\mathbf{k}}^{+} \rangle} n_{\mathbf{k}} + \frac{\langle \sigma_{\mathbf{k}}^{L} T_{\mathbf{k}}^{+} \rangle}{\langle T_{\mathbf{k}} T_{\mathbf{k}}^{+} \rangle} T_{\mathbf{k}}$$

with

$$\langle P_{\mathbf{k}} P_{\mathbf{k}}^{+} \rangle = N(\beta^2 S(k))^{-1} + \frac{m^2}{k^2} \frac{|\langle \dot{J}_{\mathbf{k}} T_{\mathbf{k}}^{+} \rangle|^2}{\langle T_{\mathbf{k}} T_{\mathbf{k}}^{+} \rangle}.$$

Entropy

$$S_{\mathbf{k}} = \left(E_{\mathbf{k}} - \frac{\langle E_{\mathbf{k}} \sigma_{\mathbf{k}}^{L+} \rangle}{\langle \sigma_{\mathbf{k}}^{L} n_{\mathbf{k}}^{+} \rangle} n_{\mathbf{k}} \right) T^{-1}$$

with

$$T \dot{S}_{\mathbf{k}} + i\mathbf{k} \cdot \hat{\mathbf{q}}_{\mathbf{k}} = 0$$

and

$$\langle S_{\mathbf{k}} S_{\mathbf{k}}^{+} \rangle = \Omega k_B C_p(k), \qquad \langle S_{\mathbf{k}} P_{\mathbf{k}}^{+} \rangle = 0.$$

The column matrix A in the generalized Langevin equation is taken to include the basic conserved variables $n_{\mathbf{k}}$, $J_{\mathbf{k}}$ $(\equiv J_{\mathbf{k}}^{L})$ and $E_{\mathbf{k}}$. By including these variables in A we ensure the correct, hydrodynamic, properties in the long-wavelength limit. It is clearly convenient to work with variables that are uncorrelated since this will simplify the matrix algebra. The variables $n_{\mathbf{k}}$ and $J_{\mathbf{k}}$ are, obviously, uncorrelated, or orthogonal, and if we use $T_{\mathbf{k}}$, instead of $E_{\mathbf{k}}$, we have three mutually orthogonal variables. With the choice,

$$A = \{n_{\mathbf{k}}, J_{\mathbf{k}}, T_{\mathbf{k}}\} \tag{3.183}$$

the square matrix $\langle AA^{+} \rangle$ is diagonal with elements given by (3.177), (3.179) and

$$\langle |T_{\mathbf{k}}|^2 \rangle = k_B T^2 / \Omega C_v(k). \tag{3.184}$$

ISBN 0-8053-6610-5, 0-8053-6611-3 (pbk.)

The elements of the random force which determine the memory function matrix, Eqs. (3.25) and (3.30), are proportional to the fluctuating components of the fluxes whose autocorrelation functions determine the transport coefficients, Eqs. (3.180)–(3.182). The fluctuating components of the fluxes are, recall, the components of the fluxes that are orthogonal to the conserved variables, i.e. A. The three elements of the random force column matrix are

$$f_1 = 0, \qquad f_2 = -(ik/m)\hat{\sigma}_{\mathrm{k}}, \qquad f_3 = -(ik/\Omega C_v(k))\hat{q}_{\mathrm{k}}, \tag{3.185}$$

and the elements of the memory function matrix $\tilde{M}$ are

$$\tilde{\theta}_2 = \langle |J_{\mathrm{k}}|^2\rangle^{-1} \int_0^\infty dt \exp(-st)\langle f_2(t) f_2^+\rangle, \tag{3.186}$$

$$\tilde{\theta}_3 = \langle |T_{\mathrm{k}}|^2\rangle^{-1} \int_0^\infty dt \exp(-st)\langle f_3(t) f_3^+\rangle, \tag{3.187}$$

$$\tilde{\theta}_{23} = \langle |T_{\mathrm{k}}|^2\rangle^{-1} \int_0^\infty dt \exp(-st)\langle f_2(t) f_3^+\rangle, \tag{3.188}$$

$$\tilde{\theta}_{32} = \langle |J_{\mathrm{k}}|^2\rangle^{-1} \int_0^\infty dt \exp(-st)\langle f_3(t) f_2^+\rangle. \tag{3.189}$$

The time development of the random force contains the modified Liouville operator that includes the projector operator. It was shown in the preceding section that the effect of the projection operator vanishes as $k \to 0$ so that in the limit $k \to 0$, $s \to 0$ the elements $\tilde{\theta}_2$ and $\tilde{\theta}_3$, Eqs. (3.186) and (3.187), are proportional to the transport coefficients $\eta_B + 4\eta/3$ and λ, respectively. In the limit $k \to 0$, $\tilde{\theta}_{23}$ and $\tilde{\theta}_{32}$ are proportional to a power of k higher than the k^2 factor in $\tilde{\theta}_2$ and $\tilde{\theta}_3$ and they are therefore neglected in the long-wavelength limit [Ernst and Dorfman (1975)].

When the various results are assembled, the matrix

$$\tilde{F}(k, s) = \int_0^\infty dt \exp(-st)\langle A(t)A^+\rangle\cdot\langle AA^+\rangle^{-1} \tag{3.190}$$

is found to be given by the result,

$$\tilde{F}(k, s)^{-1} = \begin{pmatrix} s & -\langle \dot{n}_{\mathrm{k}} J_{\mathrm{k}}^+\rangle/\langle J_{\mathrm{k}} J_{\mathrm{k}}^+\rangle & 0 \\ \langle J_{\mathrm{k}} \dot{n}_{\mathrm{k}}^+\rangle/\langle n_{\mathrm{k}} n_{\mathrm{k}}^+\rangle & \tilde{\theta}_2 + s & \tilde{\theta}_{23} - \langle \dot{J}_{\mathrm{k}} T_{\mathrm{k}}^+\rangle/\langle T_{\mathrm{k}} T_{\mathrm{k}}^+\rangle \\ 0 & \tilde{\theta}_{32} + \langle T_{\mathrm{k}} \dot{J}_{\mathrm{k}}^+\rangle/\langle J_{\mathrm{k}} J_{\mathrm{k}}^+\rangle & \tilde{\theta}_3 + s \end{pmatrix}. \tag{3.191}$$

ISBN 0-8053-6610-5, 0-8053-6611-3 (pbk.)

If we define a memory function (not a matrix)

$$\tilde{M}(k, s) = \tilde{\theta}_2 - \{\tilde{\theta}_{23} - \langle \dot{J}_{\mathbf{k}} T^+_{\mathbf{k}} \rangle / \langle T_{\mathbf{k}} T^+_{\mathbf{k}} \rangle\} \{\tilde{\theta}_{32} + \langle T_{\mathbf{k}} \dot{J}^+_{\mathbf{k}} \rangle / \langle J_{\mathbf{k}} J^+_{\mathbf{k}} \rangle\} (\tilde{\theta}_3 + s)^{-1} \quad (3.192)$$

then we find from (3.191) the result,

$$\tilde{F}_1(k, s) = \int_0^\infty dt \exp(-st) \langle n_{\mathbf{k}}(t) n^+_{\mathbf{k}} \rangle / \langle n_{\mathbf{k}} n^+_{\mathbf{k}} \rangle$$

$$= (s + \omega_0^2(k)/\{s + \tilde{M}(k, s)\})^{-1}. \quad (3.193)$$

The results (3.191)–(3.193) have a structure that is identical with that of the corresponding quantities for the paramagnet discussed in Sec. 3.3. However, we shall find the long-wavelength properties of the paramagnet and liquid quite different—in a paramagnet, the magnetization undergoes a diffusive motion whereas in a liquid, fluctuations in the number density excite a collective mode, namely, ordinary sound. The origin of this difference is the difference between the long-wavelength forms of the element $\tilde{\theta}_2$ for the two models, which in turn hinges on the marked differences in the wavevector dependence of $(\dot{M}_{\mathbf{k}}, \dot{M}^+_{\mathbf{k}})$ and $\langle |J_{\mathbf{k}}|^2 \rangle$. For the present case of liquid, the long-wavelength and small-frequency limits of the various components of the memory function are easily shown to be [Copley and Lovesey (1975), and Hansen and McDonald (1976)]

$$\tilde{\theta}_2 \rightarrow k^2(\eta_B + \tfrac{4}{3}\eta)/mn, \quad (3.194)$$

$$\tilde{\theta}_3 \rightarrow \lambda k^2 / C_v(0) \quad (3.195)$$

and

$$|\langle \dot{J}_{\mathbf{k}} T^+_{\mathbf{k}} \rangle|^2 / \langle |J_{\mathbf{k}}|^2 \rangle \langle |T_{\mathbf{k}}|^2 \rangle \rightarrow k^2 v_t^2 (\gamma - 1), \quad (3.196)$$

where

$$\gamma = C_p(0)/C_v(0) \quad (3.197)$$

and v_t is the isothermal sound velocity,

$$v_t^2 = B_T/mn = (m\beta S(0))^{-1}. \quad (3.198)$$

The factor k^2 is (3.194) is not found in the corresponding result for the paramagnet, Eq. (3.157), and, in consequence, we have sound waves in a liquid and diffusion in a paramagnet.

If we set the denominator of $\tilde{F}_1(k, s)$ equal to zero, with $s = i\omega$, and use the results (3.194)–(3.196), and solve for the poles of the resulting cubic equation in ω to lowest in k^2 we obtain

$$\omega = ik^2\lambda / C_p(0) \quad (3.199)$$

ISBN 0-8053-6610-5, 0-8053-6611-3 (pbk.)

and

$$\omega = \pm k\gamma^{1/2}v_t + \tfrac{1}{2}ik^2\left\{\left(\frac{\eta_B + \frac{4}{3}\eta}{mn}\right) + \lambda\left(\frac{1}{C_v(0)} - \frac{1}{C_p(0)}\right)\right\}. \tag{3.200}$$

The first of these poles (3.199) represents nonpropagating entropy fluctuations which result in the Rayleigh line in the light scattering spectrum from a simple liquid [Stanley (1971)]. As the liquid-gas phase transition is approached, $C_p(0)$ diverges much more strongly than λ, and so the ratio $\lambda/C_p(0)$ tends to zero. The poles (3.200) represent the propagation of ordinary sound with a damping that is proportional to k^2, and contains contributions from both viscosity and thermal conduction. The results (3.199) and (3.200) agree with the corresponding results derived from linearized hydrodynamic equations, and a complete exposition of the calculation is given by Landau and Lifshitz (1963). The behavior of the density fluctuation spectrum in the vicinity of the liquid-gas phase transition is discussed by Stanley (1971) and Swinney and Henry (1973), and Gitterman (1978).

A transport coefficient can be obtained from the limiting form of the Fourier transform of the correlation function of its associated variable. An example of this type of relation is given in (3.160). Here we record the result for the longitudinal viscosity, whose associated variable is the longitudinal particle current $J_{\mathbf{k}}$. The appropriate correlation function is given by the (2, 2) matrix element of (3.190), which we denote by $\tilde{F}_2(k, s)$. We then have

$$\eta_B + \tfrac{4}{3}\eta = nm \lim_{\omega \to 0} \omega^2 \lim_{k \to 0} k^{-2} \operatorname{Re} \tilde{F}_2(k, i\omega). \tag{3.201}$$

This result is easily verified with the aid of the results (3.193)–(3.196).

The preceding calculation confirms that, for the case of a compressible liquid, the correct long-wavelength behavior of the autocorrelation functions is obtained from the generalized Langevin equation when the initial variable comprises the basic conserved variables. Looking back at our calculation we see that a step in establishing the connection between long-wavelength phenomena derived from linearized continuum equations-of-motion [Landau and Lifshitz (1963)] and the generalized Langevin equation is that for $k \to 0$ the projection operator does not modify the temporal development of the autocorrelation functions formed from the nonconserved parts of the fluxes, so that the transport coefficients are determined by the same equations in both approaches. This result concerning the role of the projection operator as $k \to 0$ is discussed in Sec. 3.4 and an explicit calculation for a liquid is given, in detail, by Ernst and Dorfman (1975).

Short-wavelength density fluctuations in liquids can be studied by neutron scattering or computer simulation experiments, and Eq. (3.191) provides a natural starting point for the interpretation of the data. The exper-

ISBN 0-8053-6610-5, 0-8053-6611-3 (pbk.)

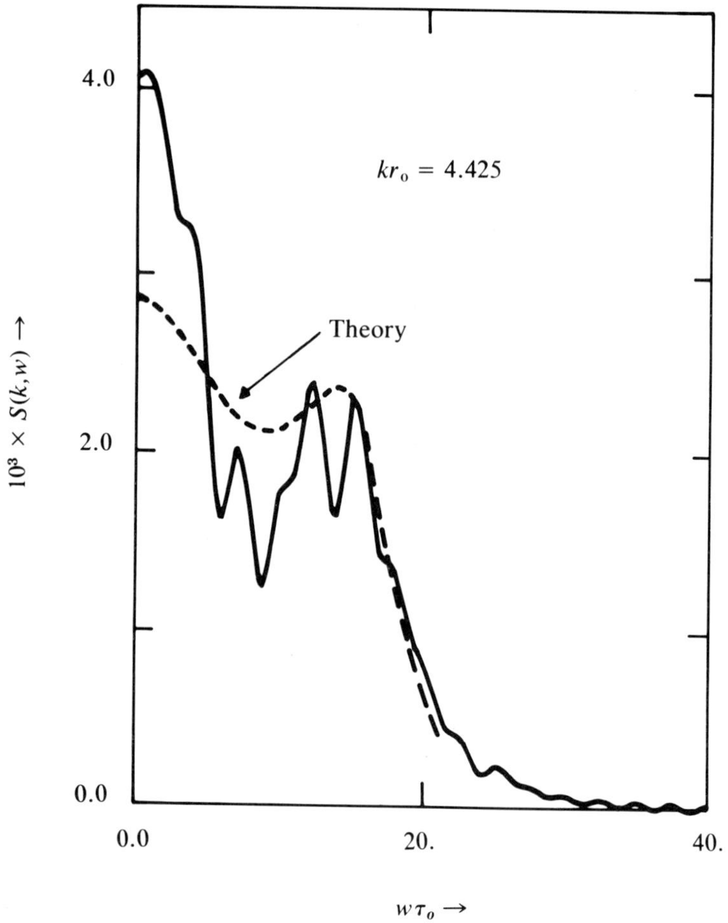

Fig. (3.2) The scattering function (3.202) is shown for $k = 4.425/r_0$ which is comparable to the position of the main peak in $S(k)$ that occurs at $k \sim 2\pi/r_0$. The frequency is given in reduced units, where the time $\tau_0 = (mr_0^2/\epsilon)^{1/2}$ and m is the particle mass. [After Lewis and Lovesey (1978).]

iments show that some liquids, in the vicinity of their triple-points, possess sufficient strongly correlated motion for them to support short-wavelength collective density oscillations [Copley and Lovesey (1975)]. One approach to the interpretation of the data on short-wavelength collective modes is to use the same short-time approximations to the memory function in (3.191) as we used in the preceding discussion of collective oscillations of the magnetization density in a paramagnet.

ISBN 0-8053-6610-5, 0-8053-6611-3 (pbk.)

In Fig. (3.2) we show some data for the scattering function

$$S(k, \omega) = \frac{1}{2\pi} \int_{-\infty}^{\infty} dt \exp(-i\omega t) \langle n_{\mathbf{k}}^{+} n_{\mathbf{k}}(t) \rangle \tag{3.202}$$

obtained in a computer simulation of a liquid in which pairs of particles interact through a potential [Lewis and Lovesey (1978)]

$$u(r) = \epsilon\{(r_0/r)^{6}2 - (r_0/r)^{4}3\},$$

where ϵ and r_0 are parameters. The quantity plotted in Fig. (3.2) is $S(k, \omega)$ for $k = 4.425/r_0$ as a function of frequency, and the temperature and density of the liquid system are $T = 0.587\epsilon$ and $n = 1.397/r_0^3$, respectively. The oscillations in the data are due mainly to numerical errors introduced in performing the temporal Fourier transform in (3.202). There is, however, clear evidence of a separate peak at a nonzero frequency which results from a collective oscillation of the number density. The dashed curve in Fig. (3.2) is obtained from a theory based on a short-time approximation to the memory function in (3.191), and it is seen to give a good account of the collective mode contribution to the scattering function.

3.6 HYDRODYNAMIC SPIN WAVES

At the critical temperature, which is determined largely by the magnitude of exchange coupling parameters, a three-dimensional magnet undergoes a spontaneous change of state from a paramagnetic to an ordered magnetic state. The phase transition is heralded by a dramatic increase in the isothermal susceptibility at wavevectors that characterize the spatial arrangement of the spins in the ordered state. The physical origin of the increase in the susceptibility in the vicinity of the critical temperature is that in changing from a paramagnet, in which the average moment is zero, to an ordered ferromagnet the magnetization in an isolated volume of the magnet undergoes large fluctuations.

The presence of long-range order leads to a number of fundamental changes in the dynamic properties of a magnet compared to those of a paramagnetic system discussed in Sec. 3.4.* For example, the energy density and magnetization are correlated in the ordered state even in the absence of an external field and the magnetization undergoes a collective motion for all wavevectors. Other differences between the dynamic properties of the paramagnetic and ordered state will show themselves in the subsequent discussion.

* An analogous situation for liquids is to compare the properties of normal and superfluid liquid states; see, for example, Forster (1975) and references therein. The hydrodynamic properties of superionic conductors have been investigated by Zeyher (1978).

ISBN 0-8053-6610-5, 0-8053-6611-3 (pbk.)

We shall focus attention in this section on the properties of the collective excitations in the ordered state, called spin waves, or magnons. Our discussion is restricted to a ferromagnetically ordered magnet, in which individual average spins are aligned in the same direction, and long-wavelength properties. Because the long-wavelength properties of a magnet do not depend on the details of the exchange forces between the spins we shall not need to specify a Hamiltonian.

The discussion given here complements the discussion of nonlinear spin dynamics given in Sec. 5.4, which also contains a discussion of the elementary properties of ferromagnetic spin waves derived from a linearized equation-of-motion. In what follows it will be useful to refer occasionally to results obtained in Sec. 5.4 to illustrate and substantiate the results derived from more general arguments.

Let the axis of quantization define the z-component of a Cartesian coordinate system. We shall assume that the interaction between the spins is isotropic, and that the total spin is a constant-of-motion. Spatial Fourier components of the spin*, $S_{\mathbf{k}}^{\alpha}$, are defined in accord with the definitions used in Sec. (5.4).

The conserved variables are the three components of the spin, $S_{\mathbf{k}}^{\alpha}$, with $\alpha = x$, y and z, and the energy density. Because of the spatial symmetry of the ordered state the transverse spin components, $S_{\mathbf{k}}^{x}$ and $S_{\mathbf{k}}^{y}$ are uncorrelated with the energy density and $S_{\mathbf{k}}^{z}$. In consequence, spin wave excitations are studied by forming a generalized Langevin equation for a two-component column matrix comprising the transverse spin components.

As in discussions given in previous sections of this chapter, we begin by calculating the static quantities that enter the frequency matrix (3.29) and the random force matrix (3.25). The transverse susceptibility $\chi_{\perp}(k)$ is defined by the equation

$$N\chi_{\perp}(k) = (S_{\mathbf{k}}^{x}, S_{-\mathbf{k}}^{x}) = (S_{\mathbf{k}}^{y}, S_{-\mathbf{k}}^{y}), \tag{3.203}$$

where the factor of the total number of spins, N, arises because of the definition of spatial Fourier transforms. A characteristic feature of the ordered state is that $\chi_{\perp}(k)$ diverges in the long-wavelength limit for isotropic spin interactions; in fact, it can be reasonably argued that the long-wavelength divergence of $\chi_{\perp}(k)$ characterizes the ordered, or broken-symmetry state. In order to find the precise nature of the divergence we appeal first to results for linear spin waves derived in Sec. 5.4. From the results obtained there we easily derive the expression, valid for linear spin waves,

$$(S_{\mathbf{k}}^{x}(t), S_{-\mathbf{k}}^{x}) = NS\cos(\omega_k t)/\omega_k, \tag{3.204}$$

* A word on notation: we can equally well speak of the spin density ($S_{\mathbf{k}}^{\alpha}$) or the magnetization density ($M_{\mathbf{k}}^{\alpha}$) as we did in Sec. 3.4. However, it is convenient in the present discussion to use spin variables, and reserve M for the average spin moment, i.e. $M = \langle S^z \rangle$.

ISBN 0-8053-6610-5, 0-8053-6611-3 (pbk.)

where S is the magnitude of the spins, and ω_k is the spin wave dispersion (5.110). Hence, from (3.203) and (3.204),

$$\chi_\perp(k) = S/\omega_k \tag{3.205}$$

and since $\omega_k \propto k^2$ for long wavelengths, Eq. (5.111), $\chi_\perp(k)$ diverges as k^{-2} as $k \to 0$ in the spin wave approximation. An alternative argument utilizes the Schwartz inequality for variables A and B,

$$(A, A^+)(B, B^+) \geq |(A, B^+)|^2. \tag{3.206}$$

If we choose $A = S_{\mathbf{k}}^x$ and $B = \dot{S}_{\mathbf{k}}^y$, then we obtain

$$(S_{\mathbf{k}}^x, S_{-\mathbf{k}}^x) \geq |(S_{\mathbf{k}}^x, \dot{S}_{-\mathbf{k}}^y)|^2/(\dot{S}_{\mathbf{k}}^y, \dot{S}_{-\mathbf{k}}^y). \tag{3.207}$$

Now,

$$\begin{aligned}(S_{\mathbf{k}}^x, \dot{S}_{-\mathbf{k}}^y) &= i\langle[S_{\mathbf{k}}^x, S_{-\mathbf{k}}^y]\rangle \\ &= i \sum_{\mathbf{lm}} \exp\{i\mathbf{k}\cdot(\mathbf{l} - \mathbf{m})\} \langle[S_{\mathbf{l}}^x, S_{\mathbf{m}}^y]\rangle \\ &= -NM, \end{aligned}\tag{3.208}$$

where M is the average spin moment. It is also straightforward to verify that

$$(\dot{S}_{\mathbf{k}}^y, \dot{S}_{-\mathbf{k}}^y) = \langle[[S_{\mathbf{k}}^y, \mathcal{H}], S_{-\mathbf{k}}^y]\rangle \tag{3.209}$$

where $\mathcal{H}$ is the Hamiltonian. Because the total y-component of the spin commutes with $\mathcal{H}$, the right-hand side of (3.209) vanishes as $\mathbf{k} \to 0$, and inversion symmetry, Eq. (3.8), requires that it vanishes as k^2, or some higher power of k^2. We therefore find that, in the long-wavelength limit,

$$(\dot{S}_{\mathbf{k}}^y, \dot{S}_{-\mathbf{k}}^y) = N\rho_0 k^2. \tag{3.210}$$

The temperature dependence of the quantity ρ_0 is determined by the pair correlation functions $\langle S_0^x S_{\mathbf{l}}^x \rangle$ and $\langle S_0^z S_{\mathbf{l}}^z \rangle$ where the maximum value of the position vector $\mathbf{l}$ is determined by the range of the exchange coupling. Using (3.208) and (3.210) in the inequality (3.207) we conclude that, in the long-wavelength limit,

$$\chi_\perp(k) \geq (M^2/\rho_0 k^2), \tag{3.211}$$

which is consistent with the previous result based on linear spin wave theory. In comparing (3.205) and (3.211) recall that, at absolute zero $M = \langle S^z \rangle = S$. In the subsequent discussion we shall write the long-wavelength limit of the transverse susceptibility in a form that is suggested by (3.211), namely,

$$\chi_\perp(k) = (M^2/\rho k^2). \tag{3.212}$$

ISBN 0-8053-6610-5, 0-8053-6611-3 (pbk.)

The only nonzero elements of $(\dot{A}, A^{+})$, namely the off-diagonal elements, are determined by (3.208). We find,

$$i\Omega = (\dot{A}, A^{+})\cdot(A, A^{+})^{-1}$$
$$= \begin{pmatrix} 0 & M/\chi_{\perp} \\ -M/\chi_{\perp} & 0 \end{pmatrix}, \qquad (3.213)$$

and the two elements of the random force

$$f = \dot{A} - i\Omega\cdot A$$

are

$$f_1 = \dot{S}_{\mathrm{k}}^x - (M/\chi_{\perp})S_{\mathrm{k}}^y \qquad (3.214)$$

and

$$f_2 = \dot{S}_{\mathrm{k}}^y + (M/\chi_{\perp})S_{\mathrm{k}}^x.$$

Because S_{k}^x and S_{k}^y satisfy conservation equations, the autocorrelation functions $(f_1(t), f_1^{\dagger}) = (f_2(t), f_2^{\dagger})$ are proportional to k^2. We define a transport coefficient ξ by the equation,

$$\xi = \int_0^{\infty} dt \lim_{k\to 0} k^{-2}(f_1(t), f_1^{\dagger}). \qquad (3.215)$$

The off-diagonal matrix elements of the memory function, $(f_1(t), f_2^{\dagger})$ and $(f_2(t), f_1^{\dagger})$ are proportional to k^2 or some higher power of k^2 in the long-wavelength limit. They do not therefore modify our results for the dispersion and damping of the spin waves to leading order in k^2, and they are omitted in the subsequent discussion.

The Laplace transform of the matrix of normalized relaxation functions formed with $A = \{S_{\mathrm{k}}^x, S_{\mathrm{k}}^y\}$ is given by,

$$\{s - i\Omega + \tilde{M}(s)\}^{-1} \qquad (3.216)$$

where $\tilde{M}(s)$ is, as before, the Laplace transform of the memory function matrix, (3.30). The (1, 1) element of (3.216) gives the Laplace transform of $(S_{\mathrm{k}}^x(t), S_{-\mathrm{k}}^x)$, and using the foregoing results for Ω and $\tilde{M}(s)$ we find that in the limit $k \to 0$, and $s \to 0$,

$$\frac{1}{N}\int_0^{\infty} dt \exp(-st)(S_{\mathrm{k}}^x(t), S_{-\mathrm{k}}^x)$$
$$= \frac{\chi_{\perp}(k)\{s + k^2\xi/\chi_{\perp}(k)\}}{\{s + k^2\xi/\chi_{\perp}(k)\}^2 + \{M/\chi_{\perp}(k)\}^2}. \qquad (3.217)$$

Setting $s = iE_k$ in the denominator of (3.217) and equating it to zero we

ISBN 0-8053-6610-5, 0-8053-6611-3 (pbk.)

find that,

$$E_k = \pm\{M/\chi_\perp(k)\} + ik^2\xi/\chi_\perp(k). \tag{3.218}$$

We conclude from this result that the spin wave dispersion

$$\omega_k = \rho k^2/M, \tag{3.219}$$

in the long-wavelength limit. Two features of this result merit explicit mention. First, the wavevector dependence is precisely the same as we find in Sec. 5.4 in the limit $k \to 0$. Secondly, the essential ingredients in the derivation given here stem directly from the presence of long-range order, which means that $M \neq 0$ and $\chi_\perp(k) \propto k^{-2}$ as $k \to 0$. This derivation contrasts with the construction of linear spin wave theory where a spin wave is a single unit of spin deviation that propagates coherently through the ordered magnet.

The damping of a spin wave in the hydrodynamic regime is seen from (3.218) to vary with the wavevector as k^4, and the ratio of the damping to the dispersion vanishes as $k \to 0$. The same wavevector dependence for the damping is obtained from the spin wave collisional self-energy, (5.141), when it is evaluated in the limit $\omega_k \ll k_B T$ [Harris (1968) and (1969)]. In the opposite limit we found, in Sec. 5.4, that the damping varies with wavevector as k^3.

The damping of long-wavelength spin waves in ferromagnetic EuO have been measured by Dietrich *et al.* (1976) and compared with the theoretical predictions. The hydrodynamic regime is defined by $k \ll \kappa$, where κ is the inverse of the correlation length. In a series of inelastic neutron scattering experiments on EuO at a temperature 5% below the critical temperature, Dietrich *et al.* measured the damping of spin waves with wavevectors in the range $0.12 \leq k \leq 0.25$ Å^{-1}, and for which $0.8 \leq (\kappa/k) \leq 1.66$. At the smallest wavevectors, and with $(\kappa/k) > 1$, the width was found to vary as k^4, but on going to the shorter wavelengths the width increases less rapidly than a fourth-power law predicts. The collisional self-energy, (5.141), evaluated in the long-wavelength limit predicts that the line width varies with k as $k^4 \ln^2(\text{const}/k^2)$. While the logarithm term gives a negligible correction to the k^4 dependence for $\kappa \gg k$, it is significant at shorter wavelengths and it reduces the value predicted by a k^4 dependence in accord with the measured values of the line widths.

ISBN 0-8053-6610-5, 0-8053-6611-3 (pbk.)

CHAPTER IV

RENORMALIZATION GROUP

The use of field-theoretic methods in statistical physics is firmly established, and Feynman diagram perturbation schemes and causal Green's functions, for example, are commonplace in the literature. One of the first significant uses of field-theoretic methods in statistical physics was made by Dyson (1956) in his study of the dynamic properties of the Heisenberg ferromagnet, a problem we treat by a different method in Chap. V. It has not been an entirely one-way process of exploiting advances in field theory in statistical physics since, for example, the concept of broken-symmetry in models of superconductivity is a cornerstone of current unified models of electromagnetic and weak interactions* [Taylor (1976) and Weinberg (1977)].

A recent example of the interplay between developments of field theory and statistical physics is the construction and use of quantitative renormalization group methods. These methods, which exploit the connection between renormalizability of perturbation theory for a model and a scale transformation, were introduced in field theory in the 1950s [for a review see, for example, Bogoliubov and Shirkov (1959)] but it was not until 1970 that the methods became truly quantitative. A method introduced by Wilson [for a review see Wilson and Kogut (1974)] has been very successful in elucidating a number of problems in statistical physics, including the Kondo problem, and phase transitions and critical phenomena; Ma (1976), Brézin *et al.* (1976), Hohenberg and Halperin (1977), and Wallace and Zia (1978).

The impact of renormalization group methods on problems in statistical physics has been so great that an introduction to it is required in any modern course on statistical physics. The application of the methods to nonequilibrium processes began some time after they were successfully used to study

* A quantum field theory of crystals, based on the spontaneous breakdown of translational symmetry and gapless Goldstone Bosons (acoustic phonons), is reviewed by Wadati (1979).

Stephen Lovesey, Condensed Matter Physics: Dynamic Correlations

ISBN 0-8053-6610-5, 0-8053-6611-3 (pbk.)

equilibrium processes. As with other topics covered in this book, we shall not aim to give a definitive account of renormalization group methods in nonequilibrium statistical physics. At the time of writing, no such account is available, and the review articles referenced in the preceding paragraph, apart from the one by Hohenberg and Halperin (1977), are concerned primarily with equilibrium properties in statistical physics. Moreover, all the articles except the one by Brézin *et al.* (1976) concentrate on Wilson's method. The introduction given here and the work by Amit (1978) and De Dominicis and Peliti (1978), is based very much on field-theoretic procedures. In particular, our account is based on ideas put forward by 't Hooft (1973) which is a particularly elegant example of what is called, for reasons that will be explained, a mass-independent renormalization method. This method, while a less physically intuitive formulation than Wilson's, offers some real advantages in actual computations because it exploits Feynman graph techniques. 't Hooft's paper is exceptionally readable, and the reader is urged to consult it. An excellent introduction to recent developments in field theory, including the method of 't Hooft and the Landau-Ginzburg model discussed in Sec. 4.6, can be found in the text book by Nash (1978).

For the most part we shall be concerned with one particular problem, namely, the incompressible Navier-Stokes liquid introduced in Sec. 2.3. The hydrodynamic properties of the incompressible Navier-Stokes liquid have been studied with Wilson's method by Forster *et al.* (1977), and the interested reader might like to make a comparative study of the Wilson and 't Hooft methods in the context of this problem. Wilson's method and field-theoretic methods for static quantities are reviewed and compared by Di Castro and Jona-Lasinio (1976). The last part of the chapter is concerned with some general features of renormalization group methods, and model systems more complicated than the Navier-Stokes liquid.

4.1 THE SPECIAL PROPERTIES OF HYDRODYNAMIC PHENOMENA IN THE NAVIER-STOKES LIQUID IN LESS THAN TWO DIMENSIONS

The type of problem that can be resolved by a renormalization group method is illustrated in a study of hydrodynamic properties of the incompressible Navier-Stokes liquid in a space dimension less than some critical dimensionality which is characteristic of the model. Above the critical dimensionality, properties of the liquid, such as those revealed by the velocity autocorrelation function, can be computed by standard methods. For example, in Chap. II we calculated the spatial Fourier transform of the velocity autocorrelation function in three dimensions by an equation-of-motion method, and we noted that the approximate equation can be obtained from a Feynman diagram expansion scheme by summing an infinite number of particular diagrams. In what follows we shall show that individual terms in

ISBN 0-8053-6610-5, 0-8053-6611-3 (pbk.)

the expansion diverge in the long-wavelength and long-time limit for space dimension d less than two.

This type of problem is well known in field theory and it is handled by using a renormalization program, described by, for example, Bogoliubov and Shirkov (1959). In fact, there are many different renormalization programs, and the one used here is chosen because it is ideal for the formulation of a particularly elegant renormalization group method. There are essentially two elements in the renormalization program: first, a procedure to give a well-defined meaning to nominally divergent contributions in the perturbation expansion and, secondly, a prescription to reorganize the expansion to give finite quantities. Although the entire program used here is based on recent advances made in field theory (high energy particle physics, to be precise), the reader should be aware that there are two basic differences between models tackled in field theory and statistical physics. The usual divergence of Feynman integrals encountered in field theory results from large momenta (so-called, ultra-violet divergences). This contrasts with the situation in statistical physics where divergences occur in the hydrodynamic limit (infrared divergence), and an upper limit on the magnitude of wave-vectors in continuum based models is the inverse of a typical intermolecular distance, i.e. an upper cutoff is intrinsic to the models. A second basic difference between field theory and statistical physics models is that in the former the physics of interest occurs in the critical dimension space of one time and three spatial coordinates, whereas in statistical physics, as often as not, the case of special interest, namely three spatial dimensions, differs from the critical dimensionality.

A renormalization program is essentially just a mathematical procedure. The physics behind the divergences in the perturbation expansion is revealed when the renormalization of the model is combined with a spatial, scale transformation, leading to a renormalization group from which the asymptotic properties of the model are deduced.

It can be argued that divergence of the perturbation expansion actually simplifies a calculation since the divergent behavior overwhelms all other types of behavior. Hence, the nub of the problem is, really, how to systematically identify and handle nominally divergent Feynman integrals.

Our main objective in this section is to demonstrate that the hydrodynamic properties of the incompressible Navier-Stokes liquid depend strongly on the spatial dimension of the model. We shall show, in fact, that the hydrodynamic properties of the model are fundamentally different above and below $d = 2$. The starting point of our discussion is the analysis of the velocity autocorrelation function given in Sec. 2.3.

The actual quantity calculated in Sec. 2.3 is related to the Laplace transform of the spatial Fourier transform of the velocity autocorrelation function, denoted by $\tilde{g}_k(s)$. In the absence of interactions this quantity

ISBN 0-8053-6610-5, 0-8053-6611-3 (pbk.)

takes the form,

$$\tilde{g}_k(i\omega) = (i\omega + \nu_0 k^2)^{-1}, \tag{4.1}$$

where ν_0 is the kinematic viscosity. Notice that $\tilde{g}_k(s)$ has the dimension of the inverse of frequency, ω.

The mixing of the velocity modes by the nonlinear terms in the Navier-Stokes equation change $\tilde{g}_k(i\omega)$ to the form,

$$\tilde{g}_k(i\omega) = \{i\omega + \nu_0 k^2 + \tilde{\Sigma}_k\,(i\omega)\}^{-1} = \Gamma_0^{-1}(k, \omega). \tag{4.2}$$

In this equation $\tilde{\Sigma}_k\,(i\omega)$ is the collisional self-energy. Because the velocity is a conserved variable $\tilde{\Sigma}_k\,(i\omega)$ is proportional to k^2, and it has the dimension of frequency.

The self-energy can be evaluated as an expansion in powers of χ, where χ is the susceptibility which enters the equilibrium distribution function of the velocity field $\mathbf{v}(\mathbf{r})$, namely

$$P_0 \propto \exp\left\{\frac{-1}{2\chi}\int d\mathbf{r}\;\mathbf{v}^2(\mathbf{r})\right\}. \tag{4.3}$$

For a Boltzmann distribution of velocities, χ is proportional to temperature and inversely proportional to the mass density. The expansion of the self-energy can be conveniently expressed in terms of Feynman diagrams, and the first few of these are shown in Fig. (2.2).

The lowest-order contribution is represented by the diagram depicted in Fig. (2.2a), and an explicit expression for it is given by Eq. (2.45) with $g_p(t)$ and $g_{k-p}(t)$ replaced by their noninteracting values. The geometric function $T_{\mathrm{k}}(\mathbf{p}, \mathbf{q})$ for space dimension d is,

$$\begin{aligned}2(d-1)T_{\mathrm{k}}(\mathbf{p}, \mathbf{q}) = {} & \{1 - (\hat{\mathbf{k}}\cdot\hat{\mathbf{p}})^2\}\{d - 2 + (\hat{\mathbf{k}}\cdot\hat{\mathbf{q}})^2\} \\ & + \{1 - (\hat{\mathbf{k}}\cdot\hat{\mathbf{q}})^2\}\{d - 2 + (\hat{\mathbf{k}}\cdot\hat{\mathbf{p}})^2\} \\ & + 2(\hat{\mathbf{k}}\cdot\hat{\mathbf{p}})(\hat{\mathbf{k}}\cdot\hat{\mathbf{q}})\{(\hat{\mathbf{p}}\cdot\hat{\mathbf{q}}) - (\hat{\mathbf{k}}\cdot\hat{\mathbf{p}})(\hat{\mathbf{k}}\cdot\hat{\mathbf{q}})\}.\end{aligned} \tag{4.4}$$

If we set aside angular terms in the wavevector integration for the moment and set $\mathbf{k} = 0$ in the integrand we obtain the following result for the lowest-order contribution to the self-energy in the long-wavelength limit,

$$\tilde{\Sigma}_k(s) = C\chi k^2 \int_0^\Lambda \frac{dp\; p^{d-1}}{s + 2\nu_0 p^2} + \cdots. \tag{4.5}$$

Here C is a constant, the intrinsic cutoff, Λ, is of the order of the inverse of a molecular length, and the Laplace variable $s = i\omega$. Because χ has dimension $(\text{length})^{d+2}/(\text{time})^2$, and ν_0 has the dimension $(\text{length})^2/\text{time}$, the right-hand side of the expression has the dimension of frequency, as required. Dimensional analysis plays an important role in subsequent discussions, so that it is expedient to write the integral in (4.5) in dimensionless

ISBN 0-8053-6610-5, 0-8053-6611-3 (pbk.)

form. Defining a wavevector $\theta_0 = (s/\nu_0)^{1/2}$, we find,

$$\tilde{\Sigma}_k(\theta_0) = C\left(\frac{k^2\chi}{2\nu_0}\right)\theta_0^{-\epsilon}\int_0^{\Lambda/\theta_0}\frac{dx\,x^{d-1}}{(\frac{1}{2}) + x^2} + \cdots \tag{4.6}$$

where $\epsilon = 2 - d$.

Let us now consider the $\theta_0 \to 0$ limit of (4.6), i.e. the long-time behavior. For $d > 2$ ($\epsilon < 0$) the integral diverges when this limit is taken, and the factor $\theta_0^{-\epsilon}$ vanishes. To obtain the appropriate limiting form in dimension $d = 3$, say, we rearrange the integral. Because,

$$\int_0^{\Lambda/\theta_0}\frac{dx\,x^{d-1}}{(\frac{1}{2}) + x^2} = \int_0^{\Lambda/\theta_0} dx\,x^{d-3}[1 - \tfrac{1}{2}(\tfrac{1}{2} + x^2)^{-1}]$$

we find,

$$\tilde{\Sigma}_k(\theta_0) = C\left(\frac{k^2\chi}{2\nu_0}\right)\left[-\frac{1}{\epsilon}\Lambda^{-\epsilon} - \tfrac{1}{2}\,\theta_0^{-\epsilon}\int_0^{\Lambda/\theta_0}\frac{dx\,x^{d-3}}{(\frac{1}{2}) + x^2}\right] + \cdots \tag{4.7}$$

and for $d < 4$ the integral in the second term tends to a finite value in the limit $\theta_0 \to 0$. Hence we obtain a result for the lowest-order contribution to the self-energy in the long-time limit ($\theta_0 \to 0$) which depends explicitly on the intrinsic cutoff.

We can use dimensional analysis to obtain the power of the contribution from the intrinsic cutoff in higher-order terms in the expansion of the self-energy. Because the expansion is in even powers of $\chi^{1/2}$ the mth-order term has the following functional form,

$$k^2\nu_0(\chi^{1/2}/\nu_0)^{2m}\theta_0^{-m\epsilon}, \tag{4.8}$$

and $m = 1, 2, \ldots$. From this result it follows immediately that the intrinsic cutoff can give a contribution proportional to $\Lambda^{-m\epsilon}$ to the mth-order term in the expansion of the self-energy when the limit $\theta_0 \to 0$ is taken. In consequence, the self-energy depends on Λ in a complicated manner for $2 < d < 4$ in the long-wavelength and long-time limit. Notice that we can extend the upper limit on the range of dimensionality by repeatedly factoring the integrals; each factoring increases the upper limit on the dimensionality by 2.

The preceding result for the dimensional form of the general term in the expansion of the self-energy can also be obtained by considering the structure of the mth-order Feynman diagram. By referring to Fig. (2.2) we deduce that the mth order diagram contains $2m$ vertices. The bare vertex function, defined in Eq. (2.28), has the dimension of wavevector. Remembering that a term k^2 factors out from these $2m$ vertices we conclude that the vertices contribute a term to the Feynman integral with the dimension θ_0^{2m-2}. There are m wavevector integrations to be performed and the inte-

ISBN 0-8053-6610-5, 0-8053-6611-3 (pbk.)

gration volume is therefore proportional to θ_0^{md}. Each of the $(2m - 1)$ denominators has a dimension of θ_0^2. Hence, the overall dimensionality of the mth order Feynman integral is proportional to θ_0^δ where the power

$$\delta = 2m - 2 + md - 2(2m - 1) = m(d - 2) = -m\epsilon, \tag{4.9}$$

and this agrees with our previous result based on dimensional analysis.

The process we have just been through to obtain the dimensionality of a Feynman integral is called power counting. If the power $\delta > 0$, the corresponding integral contains a term Λ^δ. For $\delta = 0$, the integral is logarithmically divergent as $\Lambda \to \infty$, and for $\delta < 0$, the corresponding integral converges in the limit $\Lambda \to \infty$. When we recall that in field theory there is no intrinsic cutoff and the limit $\Lambda \to \infty$ has to be taken, we understand the importance in this context of power counting and the special significance of the space dimension in which $\delta = 0$, namely $d = d_c$. To obtain finite quantities from the Feynman diagram expansion for $\delta > 0$ and $\Lambda \to \infty$ requires the introduction of an infinite number of parameters into the model to absorb the infinite parts of the Feynman diagrams. Such models are called unrenormalizable. A model with $\delta = 0$ is called renormalizable and finite quantities are obtained from the perturbation expansion by introducing a limited number of parameters to absorb logarithms of Λ. Models with $\delta < 0$ are called super-renormalizable.

In the model of interest here Λ is finite and we wish to study the hydrodynamic limit in which $k \to 0$ and $\theta_0 \to 0$. Setting $k = 0$ we find that for $\delta > 0$, which occurs when $\epsilon < 0$, the mth-order term in the expansion of the collisional self-energy can contain a term proportional to $\Lambda^{-m\epsilon}$ that is absorbed into the "bare" kinematic viscosity ν_0, and a frequency dependent term which is proportional to $\theta_0^{-m\epsilon}$ for $\theta_0 \to 0$. Consider now the behavior of the general term for $\delta < 0$, i.e. $\epsilon > 0$. Because $\delta < 0$ each integral converges as $\Lambda/\theta_0 \to \infty$, as is evident for $m = 1$ in (4.4), and there is no longer any dependence on the intrinsic cutoff. However, the contribution to the self-energy diverges as $\theta_0 \to 0$ since, according to (4.8), the mth term is proportional to $1/\theta_0^{m\epsilon}$. We conclude that the hydrodynamic properties of the incompressible Navier-Stokes liquid are fundamentally different for $\epsilon < 0$ and $\epsilon > 0$, i.e. above and below the critical dimension $d = d_c = 2$.

The gist of the renormalization program is as follows. For $\epsilon = 0$ dimensional analysis limits the divergences to the form

$$[\ln(\Lambda^2/\theta_0^2)]^m. \tag{4.10}$$

The "mass parameter" θ_0^2 is made a function of a wavevector parameter μ, and then the perturbation theory is valid when $\theta_0 \sim \mu$. It is, therefore, essential to study the properties of the model as a function of μ. Below the

ISBN 0-8053-6610-5, 0-8053-6611-3 (pbk.)

critical dimension $d = 2$ we can achieve the same goal by making a double expansion of the self-energy in χ and ϵ using the result,

$$(\theta_0^2/\mu^2)^\epsilon = \exp\{\epsilon \ln(\theta_0^2/\mu^2)\}$$
$$= \sum_{l=0}^{\infty} \frac{1}{l!} \epsilon^l \{\ln(\theta_0^2/\mu^2)\}^l. \qquad (4.11)$$

In view of this we expect to obtain properties of the model below the critical dimension in terms of an ϵ expansion. The parameter μ is determined ultimately by requiring that the physics of the model is not changed by its introduction.

The infrared divergence problem encountered here for $d < 2$ might be set aside by arguing that the physical properties of a model liquid are bound to be anomalous in less than two dimensions; for example, the conventional picture of self-diffusion must break down in less than two dimensions. The real motivation to resolve the problem posed here is more apparent when it is seen that the same kind of problem is faced in a variety of other systems, and for which the critical dimensionality is greater than $d = 3$. For example, if instead of being the velocity the dynamical variable was a conserved variable of a model with broken symmetry then the susceptibility would behave like $\chi_k \propto k^{-2}$ at the critical temperature and so the propagators in the expansion for the collisional self-energy would be proportional to $(\theta_0^2 + p^4)^{-1}$ instead of $(\theta_0^2 + p^2)^{-1}$. If there were no other differences with the Navier-Stokes system then it is evident that the critical dimensionality of the new model would be changed from $d = 2$ to $d = 4$. The continuum model of ferromagnet does have a different bare vertex function from the Navier-Stokes liquid, and the spin density is a conserved quantity. The integral for the first-order correction to the spin-diffusion constant which corresponds to (4.5) is, in the long-wavelength limit [Ma and Mazenko (1975)]

$$\int d\mathbf{p}\, (\hat{\mathbf{k}}\cdot\mathbf{p})^2 \{p^4(\theta_0 + 2p^4)\}^{-1}, \qquad \hat{\mathbf{k}} = \mathbf{k}/|\mathbf{k}|$$

and power counting shows that the critical dimensionality is $d_c = 6$. We conclude, then, that the infrared divergence encountered in the study of the Navier-Stokes liquid in the long-wavelength limit and $d < 2$ is not of solely academic interest, since models exist that have a critical dimensionality greater than $d = 3$.

4.2 DIMENSIONAL REGULARIZATION

A first step toward handling the divergence problem encountered in the preceding section is to give a well-defined meaning to the nominally divergent Feynman integrals. This step is called regularization. The method of regularization chosen here, called dimensional regularization, is based on

ISBN 0-8053-6610-5, 0-8053-6611-3 (pbk.)

the integral identity, valid for $\theta_0 > 0$, $\alpha > -1$, and $2\beta > (\alpha + 1)$,

$$\int_0^\infty dp\, p^\alpha(\theta_0^2 + p^2)^{-\beta}$$

$$= \tfrac{1}{2}\Gamma\left(\frac{\alpha+1}{2}\right)\Gamma(\beta - \tfrac{1}{2}(\alpha+1))/\{\Gamma(\beta)\theta_0^{2\beta-(\alpha+1)}\}, \qquad (4.12)$$

where $\Gamma(x)$ is the Gamma function. The Gamma functions are analytic functions, and therefore the integral is an analytic function of α,

Using this identity, the property $\Gamma(1 + x) = x\Gamma(x)$, and defining $\epsilon = 2 - d \geq 0$, we find that for $\theta_0 \to 0$ the integral in (4.5) is

$$\frac{1}{2}\int_0^\infty dp\, p^{d-1}(p^2 + \tfrac{1}{2}\theta_0^2)^{-1} = \frac{1}{2\epsilon}(\tfrac{1}{2}\theta_0^2)^{-\epsilon/2}\Gamma(1 - \tfrac{1}{2}\epsilon)\Gamma(1 + \tfrac{1}{2}\epsilon). \qquad (4.13)$$

The pole at $\epsilon = 0$ arises because the integral (4.5) is logarithmically divergent for $d = 2$ as $\Lambda/\theta_0 \to \infty$. Hence the divergence of a Feynman integral appears now as poles at $\epsilon = 0$. To lowest order in ϵ the right-hand side of (4.13) is

$$\frac{1}{2\epsilon}\{1 - \tfrac{1}{2}\epsilon \ln(\tfrac{1}{2}\theta_0)\} + \mathcal{O}(\epsilon).$$

The integral is seen from this to have a simple pole at $\epsilon = 0$, a constant term proportional to $\ln(\theta_0^2/2)$, and terms of order ϵ and higher. In a field theory, θ_0^2 would represent a mass. An important feature of the renormalization group method described here is that it is mass-independent, i.e. the logarithmic terms encountered in (4.13) do not enter in the discussion of the physical features of the model system.

Higher-order diagrams in the expansion of the self-energy have higher-order poles at $\epsilon = 0$, e.g. the diagrams depicted in Figs. (2.2b) and (2.2c) have poles of order ϵ^{-2}.

4.3 RENORMALIZATION

Having given a meaning to nominally divergent Feynman integrals, we can consider the problem of reorganizing the perturbation theory for the self-energy so as to obtain meaningful results for the velocity autocorrelation function in the hydrodynamic limit and $d < 2$. Because the infrared divergence problem is associated with the dimensionality of the model, a first step should be to expose explicitly the dependence of the model parameters on d. For the Navier-Stokes liquid the only parameter which depends on d is χ whose dimensionality is $(\text{length})^{d+2}/(\text{time})^2$. The perturbation expansion for the collisional self-energy is in terms of powers of $\chi^{1/2}$. In view of the dimensional structure of the general term in the expansion, Eq. (4.8), it is

ISBN 0-8053-6610-5, 0-8053-6611-3 (pbk.)

convenient to define a "bare" coupling constant λ_0 through the relation,

$$\chi^{1/2} = \nu_0 \lambda_0. \tag{4.14}$$

The dimension of λ_0 is (length)$^{-\epsilon/2}$ with $\epsilon = 2 - d$.

We shall now introduce a scale parameter μ which has the dimension of a wavevector, so that $\lambda_0 \sim \mu^{\epsilon/2}$. The scale parameter plays a vitally important role in formulating the renormalization group yet it does not appear in the final results of interest. To better understand the, perhaps at first sight, mysterious role of μ we stress that, through the introduction of μ, we aim to reorganize the perturbation theory to give finite results but since μ is arbitrary its introduction must not change the physics. Hence, ultimately, no physically significant quantity can depend on μ, and this requirement places important restrictions on the theory, as we shall see.

Consider now the calculation of $\Gamma_0(k, \omega)$ defined in (4.2) as the inverse of the Laplace and Fourier transform of the velocity autocorrelation function. The divergent behavior of the Feynman integrals appears in dimensional regularization as poles at $\epsilon = 0$. In order to obtain finite quantities we must evidently introduce functions which will absorb these poles. We are therefore led to consider dimensionless functions of the form

$$Z = 1 - \sum_{n=1} b_n/\epsilon^n \tag{4.15}$$

which are usually called multiplicative renormalization functions. This name is an appropriate one since we show that, with

$$\lambda_0 = \mu^{\epsilon/2}\lambda(\mu)/Z(\mu) \tag{4.16}$$

and

$$\theta^2 = Z\theta_0^2 \tag{4.17}$$

where λ is a renormalized dimensionless coupling constant, the function

$$\Gamma(k, \theta, \lambda) = \left(\frac{Z}{\nu_0}\right) (s + \nu_0 k^2 + \tilde{\Sigma}_k(s)) = \left(\frac{Z}{\nu_0}\right) \Gamma_0(k, \theta, \lambda), \tag{4.18}$$

can be made finite as $\epsilon \to 0$ order-by-order in λ. Moreover, the coefficients b_n in (4.15) can be made functions of λ only, i.e. b_n will not depend on θ^2, the "mass" parameter.

Consider the explicit form of the contribution to Γ of order λ^2, depicted in Diag. (2.1a). For long wavelengths $T_{\mathbf{k}}(\mathbf{p}, \mathbf{k} - \mathbf{p})$, Eq. (4.4), can be approximated by

$$T_{\mathbf{k}}(\mathbf{p}, \mathbf{p}) = \sin^2 \zeta (d - 2\sin^2 \zeta)/(d - 1), \tag{4.19}$$

where ζ is the angle between $\mathbf{k}$ and $\mathbf{p}$. Using this result, the contribution to

ISBN 0-8053-6610-5, 0-8053-6611-3 (pbk.)

Γ_0 of order λ_0^2 is,

$$\frac{k^2\chi}{(d-1)}\int d\Omega_{\rm p}\left\{\frac{\sin^2\zeta(d-2\sin^2\zeta)}{s+\nu_0p^2+\nu_0|\mathbf{k}-\mathbf{p}|^2}\right\}$$

$$\Rightarrow k^2\nu_0\frac{\mu^\epsilon\lambda^2}{Z^2(d-1)}\int d\Omega_{\rm p}\left\{\frac{\sin^2\zeta(d-2\sin^2\zeta)}{\theta_0^2+2p^2}\right\}. \tag{4.20}$$

In the second line of (4.20) we have taken the limit $\mathbf{k}\to 0$, and used (4.14) and (4.16). The d-dimensional integration over $\mathbf{p}$ in (4.20) is defined in the limit $\theta_0 \to 0$ by

$$\int d\Omega_{\rm p} = K_d^{-1}\int_0^\infty dp\, p^{d-1}\int_0^\pi d\zeta\,\sin^{d-2}\zeta \tag{4.21}$$

where

$$Kd = 2^{d-1}\pi^{(d+1)/2}\Gamma((d-1)/2).$$

Performing the angular integration in (4.20) for $d = 2$ leads to (4.20) being replaced by

$$\frac{k^2\nu_0\mu^\epsilon\lambda^2}{Z^2 8\pi}\int_0^\infty\frac{dp\,p^{d-1}}{(\theta_0^2+2p^2)}$$

$$=\frac{k^2\nu_0\lambda^2}{Z^2 16\pi\epsilon}\left(\frac{\mu^2}{2\theta_0^2}\right)^{\epsilon/2}\Gamma(1-\tfrac{1}{2}\epsilon)\Gamma(1+\tfrac{1}{2}\epsilon), \tag{4.22}$$

where the second equality follows from (4.13). Substituting (4.22) into (4.18) and taking the limit $\epsilon \to 0$, we find

$$\Gamma(k,\theta,\lambda) = k^2+\theta^2+k^2\{Z-1+(\lambda^2/Z16\pi\epsilon)+\cdots\}, \tag{4.23}$$

from which we conclude that, to lowest order in λ, Γ is finite as $\epsilon \to 0$ for

$$Z = 1-\lambda^2/16\pi\epsilon. \tag{4.24}$$

The result (4.24) is explicit proof that, to lowest order, Z depends on λ only. A general proof follows from the fact that Z is dimensionless, so

$$Z = Z(\mu^{-\epsilon/2}\lambda_0) = Z(\lambda/Z) = Z(\lambda). \tag{4.25}$$

This result implies nontrivial properties of the perturbation expansion because singular terms proportional to $\ln(\theta_0^2/\mu^2)$, which arise from the expansion of $(\theta_0^2/\mu^2)^\epsilon$, must cancel order-by-order in λ.

The general form for Γ is obtained from the expansion for $\tilde{\Sigma}_k(s)$ by making the replacements (a) $\theta_0 \to \theta$, (b) $\lambda_0 \to \mu^{\epsilon/2}\lambda$, and (c) replacing the bare propagator $(s+\nu_0k^2)^{-1}$ by the renormalized propagaor $\{Z(s+\nu k^2)\}^{-1}$. If we denote the value of the dimensionally regularized Feynman integrals

ISBN 0-8053-6610-5, 0-8053-6611-3 (pbk.)

associated with λ^{2m} by the dimensionless functions $f_m(\epsilon)$ then the general form for Γ is

$$\Gamma = k^2 + \theta^2 + k^2 \left\{ Z - 1 + \left(\frac{\lambda^2}{Z}\right) (\mu^2/\theta^2)^{\epsilon/2} f_1(\epsilon) + \left(\frac{\lambda^4}{Z^3}\right) (\mu^2/\theta^2)^{\epsilon} f_2(\epsilon) + \cdots \right\}. \tag{4.26}$$

Here $f_1(\epsilon)$ is obtained from the diagram depicted in Fig. (2.2a), and $f_2(\epsilon)$ is obtained from the sum of the diagrams depicted in Figs. (2.2b) and (2.2c). Writing

$$f_1(\epsilon) = t_{11}/\epsilon + t_{12} + \cdots$$
$$f_2(\epsilon) = t_{21}/\epsilon^2 + t_{22}/\epsilon + \cdots, \text{ etc}, \tag{4.27}$$

and demanding that Z in (4.26) absorbs all poles in Γ order-by-order in λ, we obtain relations between the coefficients t_{nm} in (4.27) and b_{nm} where in (4.15)

$$b_n = \sum_{m=1} \lambda^{2m} b_{nm}. \tag{4.28}$$

For example, $b_{11} = t_{11}$, and $b_{12} = b_{11}t_{12} + t_{22}$. A second set of relations comes from making the renormalization program "mass-independent", which requires the vanishing of all singular terms in Γ proportional to $\ln(\mu^2/\theta^2)$; this requirement leads, for example, to the result $t_{11}b_{11} + 2t_{21} = 0$. Combining these two sets of results we find $b_{nm} = 0$ for $n > m$, and, to order λ^6,

$$b_{22} = \tfrac{1}{2}b_{11}^2, \qquad b_{23} = \tfrac{5}{3}b_{11}b_{12}, \qquad b_{33} = \tfrac{1}{2}b_{11}^3. \tag{4.29}$$

Notice that these relations serve to determine the coefficients of the poles at ϵ^{-m}, $m \geq 2$, in terms of the coefficients of the simple pole. We shall see later that all quantities of interest are, in fact, determined from a knowledge of the coefficient of the simple pole, a very nontrivial result in actual computations since the calculation of Feynman integrals is often very tedious. The relations can be used as a check on specific calculations.

4.4 RENORMALIZATION GROUP

We have seen in the preceding section that a multiplicative renormalization program renders the function Γ, introduced in (4.18),

$$\Gamma = (Z/\nu_0)\Gamma_0 = \mu^2 A(k/\mu, \theta^2/\mu^2, \lambda) \tag{4.30}$$

finite as $\epsilon \to 0$. The second equality in (4.30) defines a dimensionless function A that will be useful in subsequent discussions. The function Γ remains arbitrary until we specify μ, which we do by demanding that the introduction of μ must not change the physics.

ISBN 0-8053-6610-5, 0-8053-6611-3 (pbk.)

Because Γ_0 determines the velocity autocorrelation function, from which we can calculate the diffusion constant, for example, it must not change under a rescaling of μ. If r is an arbitrary scale parameter, then this requirement on Γ_0 means that,

$$\Gamma_0(k, \theta(\mu), \lambda(\mu)) = \Gamma_0(k, \theta(r\mu), \lambda(r\mu)), \tag{4.31}$$

and, from this it follows that,

$$(d/dr)\Gamma_0(k, \theta(r\mu), \lambda(r\mu)) = 0. \tag{4.32}$$

Completing the operation of differentiation in (4.32) and then setting $r = 1$ leads to the result,

$$(\mu\partial_\mu + \beta\partial_\lambda + z\theta^2\partial_{\theta^2} - z)\mu^2 A(k/\mu, \theta^2/\mu^2, \lambda) = 0, \tag{4.33}$$

where the rate of change of λ is governed by the function

$$\beta(\lambda) = (d\lambda/d\ln\mu) \tag{4.34}$$

and

$$z(\lambda) = (d\ln Z/d\ln\mu) = (d\ln\theta/d\ln\mu). \tag{4.35}$$

A result like (4.33) was first derived, by quite different methods, by Callan (1970) and Symanzik (1970), (1971). The functions β and z are usually called the Gell-Mann-Low and anomalous dimension functions, respectively. Equation (4.33) differs in one important aspect from the original equations discussed by Callan and Symanzik in that it is a homogeneous equation, whereas the original equations, applied to the problem in hand, are homogeneous in the asymptotic hydrodynamic limit. Equation (4.33) can be formally solved to give $A(k/\mu, \theta^2/\mu^2, \lambda)$ by the method of characteristics. However, we choose to pursue a different route starting from (4.31). Before doing this we pause to discuss the functions β and z, Eqs. (4.34) and (4.35).

There are several features of β and z which merit explicit mention. First, both functions are logarithmic derivatives of functions with respect to μ and, in consequence, they do not change on scaling μ. Also, $z(\lambda)$ is unchanged by an arbitrary rescaling of the multiplicative renormalization function. These are all results that we should expect given the arbitrary nature of μ and Z, and the position of β and z in an equation for a renormalized quantity. From the latter observation it also follows that both β and z are well behaved in the limit $\epsilon \to 0$.

For physically significant results λ is independent of μ, so we seek the solution λ^* of the equation,

$$\beta(\lambda^*) = 0. \tag{4.36}$$

Equation (4.36) defines the fixed point in the transformation $\lambda_0 \to \mu^{\epsilon/2}\lambda(\mu)/Z(\mu)$. In addition we shall require,

$$z(\lambda^*) = z^* = \text{constant}, \tag{4.37}$$

ISBN 0-8053-6610-5, 0-8053-6611-3 (pbk.)

and from (4.35) it follows that

$$Z^* = (\mu/\Lambda)^{z^*} \tag{4.38}$$

where the factor Λ^{-z^*} is a constant of integration. In general Z is a complicated function of μ [or λ, cf. Eq. (4.40)] which means that the scaling properties of Γ_0 are nontrivial. However, at the fixed point, $Z^* \propto \mu^{z^*}$, a simple scaling behavior is recovered.

The dependence of Z on the renormalized, dimensionless coupling constant λ follows from the relation

$$z(\lambda) = \beta(\lambda)(d/d\lambda) \ln Z \tag{4.39}$$

and the boundary condition $Z(\lambda = 0) = 1$, with the result

$$Z(\lambda) = \exp\left\{\int_0^\lambda d\bar{\lambda}\, z(\bar{\lambda})\beta^{-1}(\bar{\lambda})\right\} . \tag{4.40}$$

To extract the behavior of $Z(\lambda)$ in the neighborhood of the fixed point write,

$$\beta(\lambda) = \beta(\lambda^*) + (\lambda - \lambda^*)\beta'(\lambda^*) + \cdots \tag{4.41}$$

where $\beta' = (d\beta/d\lambda)$, and set $z = z^*$ in the integrand of (4.40), so that

$$Z(\lambda) \simeq |(\lambda - \lambda^*)/\lambda^*|^{z^*/\beta'(\lambda^*)}. \tag{4.42}$$

From this result it follows that the sign of $z^*/\beta'(\lambda^*)$ determines the analytic behavior of $Z(\lambda)$ in the vicinity of the fixed-point value of the coupling constant.

A feature of particular interest is the behavior of $\lambda(\mu)$ as a function of μ in the vicinity of the fixed point. Using the result (4.41) we find

$$\lambda(\mu) \simeq (\lambda - \lambda^*)(\mu/\Lambda)^{\beta'(\lambda^*)} + \lambda^*. \tag{4.43}$$

Hence, if $\beta'(\lambda^*) > 0$ the coupling constant tends to its fixed-point value as $(\mu/\Lambda) \to 0$. A fixed point with this property is said to be infrared stable.

Thus far, we have not exploited the relation (4.16) between λ and Z. Because λ_0 is independent of μ it follows that

$$\beta(\lambda) = \lambda(z - \tfrac{1}{2}\epsilon). \tag{4.44}$$

From (4.44) it follows that the equation $\beta(\lambda^*) = 0$ admits the trivial solution $\lambda^* = 0$, and the nonconventional solution $z^* = \epsilon/2$ which is of interest. Since z^* is positive, $Z^*(\mu)$ tends to zero as $\mu \to 0$.

We return now to obtaining a formal solution for Γ_0 in terms of the function $A(k/\mu, \theta^2/\mu^2, \lambda)$ and, as mentioned already, we shall proceed directly from (4.31). If we scale the external wavevector k by the arbitrary parameter r and use (4.30) then

$$\Gamma_0(rk, \theta, \lambda) = \nu_0(r\mu)^2 Z(r\mu)^{-1} A(k/\mu, \theta^2(r\mu)/r^2\mu^2, \lambda(r\mu)). \tag{4.45}$$

Choosing $\mu = k$, we study the asymptotic behavior of Γ_0 by letting $r \to 0$

ISBN 0-8053-6610-5, 0-8053-6611-3 (pbk.)

in (4.45). Assuming that the functions on the right-hand side of (4.45) are determined in the asymptotic limit by their fixed-point values we obtain, finally,

$$\lim_{k\to 0} \Gamma_0 = \nu_0 k^2 (\Lambda/k)^{\epsilon/2} A\left(1, \frac{i\omega}{\nu_0 k^2}\left(\frac{k}{\Lambda}\right)^{\epsilon/2}, \lambda^*\right). \tag{4.46}$$

This result shows that the wavevector and frequency dependent kinematic viscosity, $\tilde{Q}_k(s)$, contains a term proportional to $k^{(d-2)/2}$ in the long-wavelength limit and $d \leq 2$. The results obtained in Sec. (2.3) for $\tilde{Q}_k(i\omega)$, generalized to space dimension $d > 2$ is

$$\tilde{Q}_k(i\omega) \propto k^{d-2} f(i\omega/\nu_0 k^2). \tag{4.47}$$

The results (4.46) and (4.47) can be combined in an appealing form

$$\tilde{Q}_k(i\omega) = (C(k)/\omega_k) h(\omega/\omega_k), \tag{4.48}$$

where $h(\omega)$ is a scaling function, and ω_k is the characteristic frequency. For $d < 2$, $\omega_k \propto k^{2-z^*} = k^{2-\epsilon/2}$ and, once again, $C(k) \propto k^d$. Hence, the anomalous behavior of the kinematic viscosity that sets in below $d = 2$ occurs in the characteristic frequency and not the amplitude factor. Usually, z^* is referred to as the anomalous scaling exponent, and it is zero for $d > 2$ and $\epsilon/2$ for $d \leq 2$. Notice that (4.48) and the results for $C(k)$ and ω_k for $d \leq 2$ imply that the long-time behavior of the wavevector and time dependent viscosity is

$$Q_k(t) \sim 1/t^{(4-2\epsilon)/(4-\epsilon)}. \tag{4.49}$$

These are exact results for the Navier-Stokes liquid and they derive directly from (4.16). In most cases more than one multiplicative renormalization parameter is required and in this instance the relation equivalent to (4.16) together with (4.36) would yield an exact relation between the various anomalous scaling exponents.

To go beyond (4.46) and obtain an explicit, approximate result for the function A we first need to know λ^*. From (4.16) and (4.34),

$$\beta = \tfrac{1}{2}\epsilon\lambda(d \ln Z/d \ln \lambda - 1)^{-1}. \tag{4.50}$$

Using (4.24) for Z we obtain from (4.50),

$$\beta = \tfrac{1}{2}\epsilon\lambda(\lambda^2/8\pi\epsilon - 1) \tag{4.51}$$

and so the nonconventional fixed-point value of the renormalized coupling constant is, to lowest order,

$$\lambda^* = (8\pi\epsilon)^{1/2}. \tag{4.52}$$

Hence,

$$\beta'(\lambda^*) = \epsilon + \mathcal{O}(\epsilon^2), \tag{4.53}$$

ISBN 0-8053-6610-5, 0-8053-6611-3 (pbk.)

showing that the nonconventional fixed point is infrared stable for $d < 2$, and, from (4.42), that $Z(\lambda)$ is well behaved in the vicinity of λ^*.

Let us turn now to the approximate calculation of the function A in (4.46) which, for notational convenience, we denote by $h(\omega/\omega_k)$ as in (4.48). A useful approximation must be based on the infinite sum of a set of diagrams, leading to a self-consistent equation for $h(\omega/\omega_k)$. In Chap. II we found that the bubble diagrams could be readily summed, so we now investigate whether this approximation is consistent with the functional form (4.46). The appropriate starting point is Eq. (2.48) for $\tilde{Q}_k(i\omega)$ after changing the spatial dimension to $d \leq 2$ which merely requires the use of (4.19). Defining

$$\nu_k(i\omega) = \nu_0 + \tilde{Q}_k(i\omega) \tag{4.54}$$

we find

$$\begin{aligned}\nu_k(i\omega) - \nu_0 = \frac{\chi}{\pi^2}\int_0^{\pi} d\zeta\, \sin^2\zeta\, \cos^2\zeta \int_0^{\infty} dp\, p^{d-1} \int_{-\infty}^{\infty} \frac{du\, du'}{(2\pi)^2} \\ \times \{(i\omega + \eta - iu - iu')[iu + p^2\nu_p(iu)] \\ \times [iu' + |\mathbf{k} - \mathbf{p}|^2 \nu_{k-p}(iu')]\}^{-1}.\end{aligned} \tag{4.55}$$

The next step is to expose the divergence of $\nu_k(i\omega)$ for $k \to 0$, and this we do by setting $k = 0$ in the integrand and placing a lower limit k on the integration range. The integration over ζ can then be completed and this gives a factor of $\pi/8$. Differentiating the resulting expression with respect to k, leads to

$$\begin{aligned}\partial_k \nu_k(i\omega) = -\frac{\chi}{8\pi} k^{d-1} \int_{-\infty}^{\infty} \frac{du\, du'}{(2\pi)^2} \{(i\omega + \eta - iu - iu') \\ \times [iu + k^2\nu_k(iu)][iu' + k^2\nu_k(iu')]\}^{-1}.\end{aligned} \tag{4.56}$$

Finally, rescale ω and the integration variables u and u' by $k^{2-\epsilon/2}$, and then it follows immediately from (4.56) that, in the limit $k \to 0$,

$$\nu_k(i\omega) = -ik^{-\epsilon/2}\chi^{1/2} h(\omega/\chi^{1/2}k^{2-\epsilon/2}), \tag{4.57}$$

where the dimensionless function $h(\omega)$ is the solution of the nonlinear integral equation

$$\begin{aligned}h(\omega) = (2\pi^2\lambda^{*2})^{-1} \int_{-\infty}^{\infty} du\, du' \\ \times \{(u + u' - \omega + i\eta)[u - h(u)][u' - h(u')]\}^{-1}.\end{aligned} \tag{4.58}$$

The result (4.57) is consistent with (4.46), as required, and the function $h(\omega/\chi^{1/2}k^{2-\epsilon/2})$ is an approximation to the function $A(1, \theta^2(k)^*/k^2, \lambda^*)$ in (4.46) derived from the sum of all the bubble diagrams. From (4.58) we

ISBN 0-8053-6610-5, 0-8053-6611-3 (pbk.)

deduce the result

$$\nu_k(0) = k^{-\epsilon/2}(\chi/8\pi\epsilon)^{1/2}, \tag{4.59}$$

which should be compared to the corresponding result for $d > 2$, given in Eq. (2.53).

4.5 GENERAL FEATURES OF A MASS-INDEPENDENT RENORMALIZATION GROUP

We stress that the results obtained with the renormalization group method follow from a reorganization of the perturbation theory for the autocorrelation function, or, more precisely, for the collisional self-energy. In view of this, a first step in tackling a particular problem of interest is to establish the rules for the perturbation expansion. It will probably not surprise the reader to learn that there are several different methods for generating perturbation expansions and the choice of method is really one of convenience and individual taste. A comparative study of the methods would take us too far afield, and good reviews exist. Perturbation methods with particular relevance to problems discussed in this book are given by Kawasaki (1976), Martin *et al.* (1973), Deker and Haake (1975a,b) and Bausch *et al.* (1976).

We have commented previously that, in view of their appearance in the Callan-Symanzik equation (4.33) for the renormalized function Γ, the functions β and z, Eqs. (4.34) and (4.35), are well behaved in the limit $\epsilon \to 0$. This leads to some general relations for z and β which are valuable in specific calculations. Consider first the behavior of $z(\lambda)$. Combining (4.44) and (4.50) we obtain the relation

$$z(\lambda) = \tfrac{1}{2}\epsilon\left\{\left(\frac{dZ}{d\ln\lambda}\right)\Big/\left[\left(\frac{dZ}{d\ln\lambda}\right) - Z\right]\right\} \tag{4.60}$$

and from (4.15)

$$\frac{d}{d\lambda} Z = -\sum_{n=1} b_n'(\lambda)/\epsilon^n \tag{4.61}$$

where $b_n' = (db_n/d\lambda)$. Substituting (4.15) and (4.61) in (4.60), and requiring that terms of order ϵ^{-n} cancel, leaving z well behaved when $\epsilon \to 0$, leads to

$$z(\lambda) = \tfrac{1}{2}\lambda b_1'(\lambda) \tag{4.62}$$

and an additional series of relations between the coefficients $b_n(\lambda)$ and $b_n'(\lambda)$, the first of which is

$$b_2'(\lambda) = \lambda^2 b_1' d(b_1(\lambda)/\lambda)/d\lambda. \tag{4.63}$$

We deduce from (4.62) that the anomalous dimension function is determined completely by a knowledge of the coefficient of the simple pole in the

ISBN 0-8053-6610-5, 0-8053-6611-3 (pbk.)

multiplicative renormalization function. For the Navier-Stokes liquid we have, to lowest order, $b_1 = \lambda^2/16\pi$, and using this result we see that (4.62) is consistent with (4.44) and (4.51). From the relation (4.63) we obtain, for example,

$$b_{22} = \tfrac{1}{2}b_{11}^2 \quad \text{and} \quad b_{23} = \tfrac{5}{3}b_{11}b_{12}.$$

These results were obtained earlier, Eq. (4.29), from the requirements that Z absorbs all poles in the expansion of the renormalized function Γ in powers of λ, and the vanishing of all singular terms in Γ proportional to $\ln(\mu^2/\theta^2)$.

For the Navier-Stokes liquid there is a simple relation between β and z, Eq. (4.44), which follows directly from the relation (4.16) that is verified perturbationally. In general, however, we must write in place of (4.16)

$$\lambda_0 = \mu^{\epsilon\alpha}\{\lambda + \sum_{n=1} a_n(\lambda)/\epsilon^n\} \tag{4.64}$$

where ϵ is again the deviation from the critical dimensionality, and α is chosen to make the renormalized coupling constant dimensionless. From the requirement that β is well behaved as $\epsilon \to 0$, it follows that

$$\beta(\lambda) = \alpha\lambda\{\lambda d(a_1(\lambda)/\lambda)/d\lambda - \epsilon\}. \tag{4.65}$$

Here, again, β is determined from a knowledge of the coefficient of the simple pole in (4.64). The value of the coupling constant at the fixed point is determined from

$$\lambda^* d(a_1(\lambda)/\lambda)/d\lambda|_{\lambda=\lambda^*} = \epsilon. \tag{4.66}$$

Because $a_1(\lambda)$ is independent of ϵ, the only place that ϵ occurs in the calculation of λ^* is its appearance on the right-hand side of (4.66).

4.6 TIME-DEPENDENT LANDAU-GINZBURG MODEL

This model has been defined, and some of its properties discussed, in Sec. 2.5. In terms of the general discussion given in Sec. 2.4 the streaming velocity in the stochastic equation for the random variables $\varphi(\mathbf{r})$ is set equal to zero, and the free energy (2.95) contains a self-coupling term $\varphi^4(\mathbf{r})$ of strength g_0. The Landau-Ginzburg model is, in some sense, the opposite limiting case to models like the Navier-Stokes liquid for which the free-energy is quadratic in the dynamical variables and the nonlinear terms in the equation-of-motion come from the streaming velocity, or mode-coupling terms. The nonlinearity in the equation-of-motion for the Landau-Ginzburg model is cubic in contrast to the quadratic, mode-coupling term found for the Navier-Stokes liquid.

The Landau-Ginzburg model is usually called the φ^4 model in field theory, for an obvious reason. It has been studied extensively, and, in particular, Amit (1978) and Nash (1978) give detailed accounts of mass-

ISBN 0-8053-6610-5, 0-8053-6611-3 (pbk.)

independent renormalization for the model. In view of this we shall in the present discussion be primarily interested in drawing attention to features of the model not found in the Navier-Stokes model.

The starting point is Eq. (2.105) for the response function $\tilde{K}_k(s)$. We work with the response function rather than the autocorrelation function because the latter has a more complicated frequency dependence than $\tilde{K}_k(s)$, and the response function is the fundamental function for the description of the physics of the model. Taking L_k to be a constant ν_0 which has the dimension of (length)²/time, we define Γ_0 to be the inverse of $\tilde{K}_k(s)$ apart from a factor ν_0, so that, using (2.105) and (2.108),

$$\Gamma_0 = \theta_0^2 + m_0^2 + k^2 + g_0 \frac{1}{2\Omega} \sum_{\mathbf{p}} \chi_p + \theta_0^2 \Sigma_k(\theta_0) - \Sigma_k(0), \quad (4.67)$$

where $\theta_0^2 = i\omega/\nu_0 = s/\nu_0$ and g_0 has dimension of $(\text{length})^{d-4}$. The self-consistent equation for χ_k is

$$\chi_k^{-1} = m_0^2 + k^2 + g_0 \frac{1}{2\Omega} \sum_{\mathbf{p}} \chi_p - \Sigma_k(0). \quad (4.68)$$

The lowest-order diagrams for the self-energy in (4.67) and (4.68) are depicted in Fig. (2.3). These diagrams are of order g_0^2 and g_0^3.

A difference between the present model and the Navier-Stokes liquid is that in renormalizing the model it is not sufficient to consider just Γ_0, since higher-order correlation functions diverge in a manner that is not controlled by renormalizing Γ_0. The higher-order correlation function must, of course, be defined systematically so that between them they contain all diagrams. Using numbers to denote wavevectors, and summing over barred numbers, the self-energy can be expressed in the form

$$\Sigma_1 = \tfrac{1}{6}\gamma(1\bar{2}\bar{3}\bar{4}) G_{\bar{2}} G_{\bar{3}} G_{\bar{4}} \Gamma_4(\bar{2}\bar{3}\bar{4}1) \quad (4.69)$$

where the bare vertex

$$\gamma(1234) = (g_0/\Omega)\delta_{1+2+3+4,0} \quad (4.70)$$

and $\Gamma_4(2341)$ is a higher-order correlation function, often called a vertex function. The first few terms in the expansion of Γ_4 in terms of the bare vertex are [Deker and Haake (1975b)]

$$\begin{aligned}\Gamma_4(1234) &= \gamma(1234) + \tfrac{1}{2}\gamma(12\bar{5}\bar{6}) G_{\bar{5}} G_{\bar{6}} \gamma(\bar{5}\bar{6}34) \\ &\quad + \gamma(13\bar{5}\bar{6}) G_{\bar{5}} G_{\bar{6}} \gamma(\bar{5}\bar{6}42) + \cdots. \end{aligned} \quad (4.71)$$

Retaining the lowest-order approximation for Γ_4 gives us back the result for the self-energy given in Sec. 2.5.

In performing the renormalization program we need to expand all quantities in terms of the coupling constant g_0, and this means replacing the

ISBN 0-8053-6610-5, 0-8053-6611-3 (pbk.)

correlation functions G in (4.69) and (4.71) by their zeroth-order values. The wavevector dependence of G is the same as that for the Navier-Stokes fluid, i.e. $\tilde{G}_k(s) \propto (k^2 + m_0^2)^{-1}$. Power counting shows that the critical dimensionality is $d_c = 4$.

We affect the renormalization by introducing renormalized (dimensionless) parameters g, m^2 and θ^2 through the transformations

$$g_0 = \mu^{\epsilon}\{g + \sum_{n=1} a_n(g)\epsilon^{-n}\}, \tag{4.72}$$

$$m_0^2 = \mu^2 m^2\{1 + \sum_{n=1} b_n(g)\epsilon^{-n}\} \tag{4.73}$$

and

$$\theta_0^2 = \mu^2\theta^2\{1 + \sum_{n=1} d_n(g)\epsilon^{-n}\}. \tag{4.74}$$

It is also necessary to renormalize the random variables $\varphi \rightarrow Z^{1/2}\varphi$ where

$$Z = 1 + \sum_{n=1} c_n(g)\epsilon^{-n}. \tag{4.75}$$

The coefficients b_n, c_n and d_n are to be determined by requiring that $Z_3\Gamma_0$ is finite as $\epsilon \rightarrow 0$ order-by-order in g. The coefficients a_n are determined by the renormalization of Γ_4, Eqs. (4.69) and (4.71). The first-order contributions to the coefficients are of order g, as is clear from (4.67) and (4.68).

From the preparatory discussion the reader will be aware that the renormalization program for the Landau-Ginzburg model is more complicated than that for the Navier-Stokes liquid. It is therefore as well to take advantage of any simplification in the algebra we can find. One simplification comes from exploiting the arbitrary nature of the renormalized coupling constant. The arbitrariness is removed, of course, at the fixed point but during the course of implementing the renormalization program we can multiply it by any well-behaved function. This being so, we define the dimensionally regularized integrals with

$$\int d\Omega_{\mathbf{p}} = (2\pi)^{-d}\int d\mathbf{p}$$

$$= \{K_d\Gamma(3 - \tfrac{1}{2}d)\}^{-1}\int_0^{\pi} d\zeta\,\sin^{d-2}\zeta\int_0^{\infty} dp\,p^{d-1}, \tag{4.76}$$

where K_d is given by (4.21). The additional factor $1/\Gamma(3 - d/2)$ in the definition avoids the appearance of Euler's constant at intermediate stages of the renormalization program.

We shall consider the steps in the renormalization program to order g^2. To carry the program to this order we need χ_k to order g. Noting that the self-energy is of order g^2, we find from (4.68)

$$\chi_k^{-1} = k^2 + m^2\mu^2(1 + b_{11}g/\epsilon) + \tfrac{1}{2}\mu^{\epsilon}g\int d\Omega_{\mathbf{p}}\,(p^2 + m^2\mu^2)^{-1}, \tag{4.77}$$

ISBN 0-8053-6610-5, 0-8053-6611-3 (pbk.)

and we shall later find that $b_{11} = 1/16\pi^2$. From (4.76),

$$\int d\Omega_p\,(p^2 + m^2\mu^2)^{-1} = -m^2\mu^2 2b_{11}/\epsilon + \cdots. \tag{4.78}$$

The lowest-order collisional self-energy contribution is depicted in Fig. (2.3). An explicit expression is obtained directly from (2.103) with $G_p(t)$ replaced by its bare value and subtracting $\Sigma_k(0)$

$$\tfrac{1}{6}(\mu^\epsilon g)^2 \int d\Omega_p\,d\Omega_q\,\{\theta^2\mu^2(\theta^2\mu^2 + p^2 + |\mathbf{p} + \mathbf{q} - \mathbf{k}|^2 + q^2)^{-1} - 1\}\chi_p\chi_{p+q-k}\chi_q \tag{4.79}$$

where the χ's are given by the first two terms in (4.77). The calculation of the singular parts of (4.79) involves quite a lot of algebra and it is discussed in great detail by Collins (1974). In view of this, we give here only the required singular parts of the integral (4.79), and these are

$$\mu^2\theta^2(g^2/8\epsilon)\ln(4/3) + \mu^2 m^2(gb_{11}/\epsilon)^2 m^{-2\epsilon} + (g^2 b_{11}^2 k^2/12\epsilon) + (\mu^2 m^2 g^2 b_{11}^2/2\epsilon). \tag{4.80}$$

Notice that in this expression of order g^2 we have terms in both ϵ^{-1} and ϵ^{-2}, and a term proportional to k^2. Much the same manipulations and tricks used in obtaining (4.80) from (4.79) are required in the calculation of the singular parts of Γ_4, Eq. (4.71). There are three contributions of order g^2, and they all contain the integral

$$\tfrac{1}{2}(g\mu^\epsilon)^2 \int d\Omega_p\,\{(p^2 + \mu^2 m^2)(|\mathbf{k} + \mathbf{p}|^2 + \mu^2 m^2)\}^{-1} = (-g^2 b_{11}/\epsilon) + \cdots. \tag{4.81}$$

Using this result in (4.71), together with the transformations (4.72)(4.75), and requiring that Γ_4 is well behaved in the limit $\epsilon \to 0$, it is found that

$$a_{12} = 3b_{11}. \tag{4.82}$$

The coefficients b_n, c_n and d_n are obtained, to order g^2, from requiring that $Z_3\Gamma_0$ is finite as $\epsilon \to 0$. The contribution from the first four terms in (4.67) are obtained using (4.77) for χ_p in the term $\Sigma\chi_p$, and the sum of the singular terms is

$$\mu^2 m^2\{-(gb_{11}/\epsilon)(1 + gm^{-\epsilon}a_{12}/\epsilon) - (gb_{11}/\epsilon)^2 m^{-\epsilon} + (gb_{11}/\epsilon)^2 m^{-2\epsilon}\}. \tag{4.83}$$

In obtaining (4.83) we have used the result

$$\int d\Omega_p\,(p^2 + \mu^2 m^2)^{-2} = (2b_{11}/\epsilon) + \cdots. \tag{4.84}$$

Notice that we have retained in (4.83) factors $m^{-\epsilon}$ and $m^{-2\epsilon}$ when they

ISBN 0-8053-6610-5, 0-8053-6611-3 (pbk.)

multiply terms which contain ϵ^{-2}. These factors, when expanded in ϵ, will give simple poles in $Z_3\Gamma_0$ with coefficients proportional to ln m^2. For the renormalization program to be mass-independent these terms must cancel amongst themselves, and it may be quickly verified that this does in fact occur. Assembling all the terms, the results for the coefficients in the transformations (4.73)–(4.75) are, in addition to (4.82),

$$b_{11} = 1/16\pi^2, \qquad b_{12} = -5b_{11}^2/12, \qquad b_{22} = 2b_{11}^2,$$
$$c_{12} = -b_{11}^2/12, \qquad d_{12} = -\tfrac{1}{8}\ln(4/3) + \tfrac{1}{12}b_{11}^2. \tag{4.85}$$

The renormalization group is constructed following the steps in Sec. 4.4. By analogy with (4.30) we write Γ_0 in terms of a dimensionless function $A(k/\mu, \theta, m, g)$, namely

$$\Gamma_0 = (\mu^2/Z_3)A(k/\mu, \theta, m, g). \tag{4.86}$$

The renormalization program described in preceding paragraphs leaves the function A well behaved as $\epsilon \to 0$ order-by-order in g. Requiring that Γ_0 is unchanged under a scaling of μ leads to a Callan-Symanzik equation with coefficients

$$\beta = (dg/d\ln\mu), \qquad z_\theta = -(d\ln\theta/d\ln\mu),$$
$$z_m = -(d\ln m^2/d\ln\mu) \quad \text{and} \quad z = (d\ln Z_3/d\ln\mu). \tag{4.87}$$

The general results of Sec. 4.5 enable us to write these functions in terms of the coefficients a_1, b_1, c_1 and d_1. The explicit results are readily shown to be,

$$\beta = g\{gd(a_1/g)/dg - \epsilon\}, \tag{4.88}$$

$$z_\theta = 2 - gdd_1/dg, \tag{4.89}$$

$$z_m = 2 - gdb_1/dg \tag{4.90}$$

and, finally,

$$z = -gdc_1/dg. \tag{4.91}$$

From the result $a_1(g) = 3g^2b_{11}$, which is valid to order g^2, the fixed-point value of the coupling constant g^*, defined by $\beta(g^*) = 0$, is

$$g^* = \epsilon/3b_{11} = 16\pi^2\epsilon/3. \tag{4.92}$$

Also,

$$\beta'(g^*) = \epsilon + \mathcal{O}(\epsilon^2). \tag{4.93}$$

The results (4.92) and (4.93) are correct to first order in ϵ. The fixed-point values of the anomalous scaling functions (4.89)–(4.91) are obtained from

ISBN 0-8053-6610-5, 0-8053-6611-3 (pbk.)

(4.85) and (4.92), and the results are

$$z_\theta^* = 2 - \eta + (\tfrac{1}{2}g^*)^2 \ln(4/3) = 2 - \eta + \eta_\theta, \tag{4.94}$$

$$z_m^* = 2 - \eta - \tfrac{1}{3}\epsilon \tag{4.95}$$

and

$$z^* = (\epsilon^2/54) = \eta. \tag{4.96}$$

The second equalities in (4.94) and (4.96) define exponents η_θ and η, respectively.

If we denote the scaling function obtained from $A(1, \theta^{2*}, m^{2*}, g^*)$ by f, the results of the renormalization group lead to the following form for Γ_0,

$$\lim_{k \to 0} \Gamma_0 = k^2(\Lambda/k)^\eta f\left\{\frac{i\omega}{\nu_0\Lambda^2}\left(\frac{\Lambda}{k}\right)^{z_\theta^*}, m^2\left(\frac{\Lambda}{k}\right)^{z_m^*}\right\}. \tag{4.97}$$

The result (4.97) enables us to establish the physical significance of the exponents (4.94)–(4.96). Let us consider the situation encountered in critical phenomena. In this case we know, from Sec. 2.4, that the coefficient of φ^2 in the expansion of the free-energy vanishes at the critical temperature T_c, i.e. $m^2 = 0$ for $T = T_c$. Setting $\omega = m^2 = 0$ in (4.97) we find that, at the critical temperature,

$$\chi_k \propto k^{-2+\eta}, \tag{4.98}$$

with η given by (4.96). The result (4.98) shows that the fourth-order terms in the free-energy modify the naive wavevector dependence of χ_k. A second interesting result gives the behavior of χ_0 in the vicinity of the critical temperature. Assume that m^2 vanishes at the critical temperature like

$$m^2 \sim (T - T_c)^{2\nu} \tag{4.99}$$

and that

$$\chi_0^{-1} \sim (T - T_c)^\gamma. \tag{4.100}$$

Taking $\omega = 0$ in (4.97), together with the requirement that the right-hand side is independent of k in the limit $k \to 0$ shows that

$$\gamma = 2\nu(2 - \eta)/z_m^*, \tag{4.101}$$

with z_m^* given by (4.95). Using the mean-field value $\nu = 1/2$ we find, to first order,

$$\gamma = 1 + \tfrac{1}{6}\epsilon. \tag{4.102}$$

Finally, consider the dynamic properties of the system at the critical tem-

ISBN 0-8053-6610-5, 0-8053-6611-3 (pbk.)

perature. From (4.97) it follows that

$$\lim_{k \to 0} \lim_{\omega \to 0} \Gamma_0 \sim (i\omega)^{\bar{\gamma}} \tag{4.103}$$

where the exponent $\bar{\gamma}$ is,

$$\bar{\gamma} = 1 - \tfrac{1}{2}\eta_\theta . \tag{4.104}$$

This result is the analogue for time-dependent Landau-Ginzburg model of (4.49) for the Navier-Stokes liquid.

ISBN 0-8053-6610-5, 0-8053-6611-3 (pbk.)

CHAPTER V

CAUSAL GREEN'S FUNCTIONS

Causal Green's functions, introduced in Sec. 1.6, have proved useful in describing the dynamic properties of a diverse range of quantal models of condensed matter physics. An important step in the acceptance and development of the method was the publication of a review of the properties of causal Green's functions by Zubarev (1960).

The reader might be aware of another Green's function commonly used in statistical physics that is usually called a time-ordered Green's function. This type of Green's function is particularly well suited for the development of diagrammatic perturbation expansions [Fetter and Walecka (1971) and Doniach and Sondheimer (1974)] whereas a causal Green's function is usually calculated from its equation-of-motion. An exception is the work by Kadanoff and Baym (1962) who discuss equation-of-motion calculational procedures for time-ordered Green's functions, and develop, so-called, conserving approximation schemes which ensure that approximate Green's functions fulfill certain important conservation equations. An introduction to this approach is given by Ziman (1969).

In this chapter we shall calculate causal Green's functions for three models of interest using equation-of-motion methods. The usefulness of Green's functions is really twofold. First, they afford a very compact and convenient computational scheme. This feature is exploited in the study of mixed, or impure systems for which the equations-of-motion are linear, and a major problem is to account adequately for the configurational averaging. We illustrate this type of calculation in a study of a mass-defect in a harmonic lattice. However, we shall for the most part be concerned with a single impurity. The reader might like to compare the mass-impurity problem with the even simpler problem of a localized, structureless impurity in an electron gas, cf. Ziman (1969).

A second useful feature of causal Green's functions is the ability to

ISBN 0-8053-6610-5, 0-8053-6611-3 (pbk.)

Stephen Lovesey, Condensed Matter Physics: Dynamic Correlations

quite easily obtain approximate results for nonlinear systems which are difficult or tedious to obtain by alternative methods. This is achieved by decoupling the hierarchy of equations-of-motion for the Green's function of interest by approximating a higher-order Green's function in terms of a product of a static correlation function and lower-order Green's functions, i.e. a Green's function with fewer variables. An example of this procedure is given in Sec. 1.7 where the displacement Green's function for an atom in an anharmonic potential is obtained in the mean-field approximation. A drawback to decoupling procedures is that there is no systematic prescription for affecting the decoupling, and so a particular approximate result must be checked carefully to ensure that it makes good physical sense. However, as one might expect after nearly two decades of using causal Green's functions, several decoupling procedures are known to give results obtained by other methods, and they can be used with some confidence. The reader should nonetheless take heed of the fact that the calculation of physically sensible results for many-particle systems often entails subtle effects which can be masked in decoupling procedures.

In the following section we review the salient properties of causal Green's functions. Subsequent sections are devoted to the study of a mass impurity in a harmonic lattice, the degenerate electron gas and, finally, spin wave interactions in a ferromagnet.

5.1 PROPERTIES OF CAUSAL GREEN'S FUNCTIONS

A causal Green's function for the operators A and B is defined as in Eq. (1.64),

$$G(t) = -i\theta(t)\langle[A(t), B]\rangle = \langle\langle A(t); B\rangle\rangle \tag{5.1}$$

where $\theta(t)$ is the unit step function ($= 1$ for $t > 0$, and $= 0$ for $t \leq 0$), and the second equality defines a convenient, and widely used, notation. The Fourier transform of $G(t)$,

$$G(\omega) = \frac{1}{2\pi}\int_{-\infty}^{\infty} dt\, G(t) \exp(-i\omega t) = \langle\langle A; B\rangle\rangle \tag{5.2}$$

satisfies the equation-of-motion

$$-\omega G(\omega) = \frac{1}{2\pi}\langle[A, B]\rangle + \langle\langle[A, \mathscr{H}]; B\rangle\rangle \tag{5.3}$$

where $\mathscr{H}$ is the Hamiltonian that describes the quantal system. In Sec. 1.6 we attached a suffix ω to the double bracket notation for the Fourier transform of a Green's function but we shall not do so here in order to simplify notation. This should not cause confusion to the reader since we shall nearly always work with the Fourier transform of $G(t)$ rather than $G(t)$ itself.

ISBN 0-8053-6610-5, 0-8053-6611-3 (pbk.)

The physical import of a causal Green's function stems from its close relation to the dynamic susceptibility. If a system is perturbed by a probe of strength h_0, described by the Hamiltonian,

$$-A^+h(t) = -A^+h_0 \exp(i\omega t + \eta t), \qquad \eta \to 0 \tag{5.4}$$

then, as shown in Sec. 1.5, the change in the average value of A induced by the probe is,

$$-\int_{-\infty}^{\infty} d\bar{t}\; h(t - \bar{t})G(\bar{t}) = h_0 \exp(i\omega t)\tilde{\chi}(i\omega) \tag{5.5}$$

where $\tilde{\chi}(i\omega)$ is the dynamic susceptibility, Eq. (1.48). From this result and the definition of $G(\omega)$, Eq. (5.2), we have immediately

$$\tilde{\chi}(i\omega) = -2\pi\langle\langle A;\, A^+\rangle\rangle. \tag{5.6}$$

This is the relation referred to above. If the probe couples to a variable that undergoes a collective oscillation then the susceptibility and $G(\omega)$ will display a resonance when ω coincides with the collective mode frequency.

An additional important result relates $G(\omega)$ to the spectrum of spontaneous fluctuations in A, namely,

$$S(\omega) = \frac{1}{2\pi}\int_{-\infty}^{\infty} dt \exp(-i\omega t)\langle A^+A(t)\rangle. \tag{5.7}$$

In Chap. 1, the function $S(\omega)$ was referred to as the scattering function because of its intimate relation to the cross section for single scattering of radiation. The relation between $S(\omega)$ and the associated causal Green's function is ($\eta \to 0$),

$$G(\omega) = \langle\langle A;\, A^+\rangle\rangle = \frac{1}{2\pi}\int_{-\infty}^{\infty} \frac{du\; S(u)\{\exp(-\beta u) - 1\}}{(u - \omega + i\eta)} \tag{5.8}$$

$$= \frac{1}{2\pi^2}\int_{-\infty}^{\infty} \frac{du\; \tilde{\chi}''(iu)}{(u - \omega + i\eta)} \tag{5.9}$$

$$= -\frac{1}{\pi}\int_{-\infty}^{\infty} \frac{du\; G''(u)}{(u - \omega + i\eta)}. \tag{5.10}$$

In the last two equalities, $\tilde{\chi}''$ and G'' are the imaginary parts of the susceptibility and Green's function $G = \langle\langle A;\, A^+\rangle\rangle$, respectively. Because $S(\omega)$ is purely real, it follows from (5.8) and the identity (1.73), that,

$$S(\omega) = 2\{1 - \exp(-\omega\beta)\}^{-1}\langle\langle A;\, A^+\rangle\rangle''. \tag{5.11}$$

The imaginary part of the Green's function is calculated from the equation-of-motion (5.3) with the substitution $\omega \to \omega - i\eta$, and $\eta \to 0$, as is evident from the requirement that the integral (5.2) converges, or alternatively from (5.8).

ISBN 0-8053-6610-5, 0-8053-6611-3 (pbk.)

From the condition of detailed balance, Eq. (1.38), and (5.11) it follows that the imaginary part of the Green's function formed from A and A^+, and the corresponding $\tilde{\chi}''(i\omega)$, are odd functions of the frequency, ω.

The dispersion relation (5.10) provides a connection between the real and imaginary parts of $\langle\langle A; A^+\rangle\rangle$. A slightly more general result can be obtained by observing that $G(\omega)$ can be continued analytically into the upper half of the complex plane, and assuming that $G(\omega)$ vanishes for $|\omega| \to \infty$. Cauchy's integral theoren then gives the result ($\eta \to 0$),

$$\int_{-\infty}^{\infty} \frac{du\, G(u)}{u - \omega - i\eta} = 0. \tag{5.12}$$

If we separate the real and imaginary parts of this result we recover (5.10), and also the conjugate result for G'' and the principal part integral of the real part of G. Landau and Lifshitz (1959) give a detailed discussion of some general properties of $G(\omega)$ that can be derived using the methods of the theory of functions of a complex variable.

5.2 MASS IMPURITY IN A HARMONIC LATTICE

We consider the lattice vibrations of a model of a crystal in which the atoms interact through a term that is quadratic in the displacements of the atoms from their equilibrium positions, and where the atoms have either a mass M or M'. The force constants, i.e. the coefficients of the quadratic interaction, are assumed not to depend on the mass of the atoms.

The Hamiltonian of the pure, or unperturbed, crystal is

$$\mathcal{H}_0 = \sum_{l\alpha} \frac{1}{2M} \{p_\alpha(l)\}^2 + \frac{1}{2} \sum_{ll'} \sum_{\alpha\beta} Y_{\alpha\beta}(l, l') u_\alpha(l) u_\beta(l'). \tag{5.13}$$

Here, the equilibrium positions of the atoms are defined by lattice vectors $\mathbf{l}$, which define a single Bravais lattice, $\mathbf{u}(l)$ is the displacement of an atom from the position defined by $\mathbf{l}$, and $\mathbf{p}(l)$ is the momentum operator conjugate to the displacement. α and β denote Cartesian components of a vector. The force constants $Y_{\alpha\beta}(l, l')$ satisfy a number of important relations [Peierls (1955) and Maradudin *et al.* (1963)].

Because $Y_{\alpha\beta}(l, l')$ is the second derivative of an interatomic potential with respect to the displacements it must clearly satisfy

$$Y_{\alpha\beta}(l, l') = Y_{\beta\alpha}(l', l) \tag{5.14}$$

and can depend only on the relative positions of the relevant unit cells in the crystal, i.e.

$$Y_{\alpha\beta}(l, l') = Y_{\alpha\beta}(l - l'). \tag{5.15}$$

Also, since every atom in a Bravais lattice is a center of inversion symmetry

$$Y_{\alpha\beta}(l, l') = Y_{\alpha\beta}(l', l) = Y_{\beta\alpha}(l, l'). \tag{5.16}$$

If we remove the atoms at the sites $l = e$ and replace them with atoms

ISBN 0-8053-6610-5, 0-8053-6611-3 (pbk.)

of mass M' then the total Hamiltonian is

$$\mathcal{H} = \mathcal{H}_0 + \mathcal{H}_1 \tag{5.17}$$

where

$$\mathcal{H}_1 = \frac{1}{2}\left(\frac{1}{M'} - \frac{1}{M}\right) \sum_{e\alpha} \{p_\alpha(e)\}^2$$

$$= \frac{1}{2}\frac{\lambda}{M'} \sum_{e\alpha} \{p_\alpha(e)\}^2. \tag{5.18}$$

The second equality in (5.18) defines a perturbation parameter λ which satisfies,

$$1/\lambda = (1 - M'/M)^{-1}. \tag{5.19}$$

We study the lattice vibrations of the system described by (5.17) by constructing the equation-of-motion for the displacement Green's function

$$G_{\alpha\beta}(l, l'; \omega) = \langle\langle u_\alpha(l); u_\beta(l')\rangle\rangle. \tag{5.20}$$

In constructing the equation-of-motion from (5.3) we shall need the commutation relation

$$[u_\alpha(l), p_\beta(l')] = i\delta_{\alpha\beta}\delta_{ll'} \tag{5.21}$$

all other commutators being zero.

We first calculate the displacement Green's function for the pure crystal, and this function we denote by $P_{\alpha\beta}(l, l'; \omega)$. From the equation-of-motion

$$-\omega P_{\alpha\beta}(l, l'; \omega) = \frac{1}{2\pi}\langle[u_\alpha(l), u_\beta(l')]\rangle + \langle\langle[u_\alpha(l), \mathcal{H}_0]; u_\beta(l')\rangle\rangle$$

we find

$$\omega P_{\alpha\beta}(l, l'; \omega) = -(i/M)\langle\langle p_\alpha(l); u_\beta(l')\rangle\rangle. \tag{5.22}$$

To obtain the Green's function of the right-hand side of (5.22) we evidently need to evaluate the commutator

$$[p_\gamma(l''), \mathcal{H}_0] = \tfrac{1}{2} \sum_{ll'} \sum_{\alpha\beta} Y_{\alpha\beta}(l, l')[p_\gamma(l''), u_\alpha(l)u_\beta(l')]$$

$$= -i \sum_{l'\beta} Y_{\gamma\beta}(l'', l')u_\beta(l'). \tag{5.23}$$

Using this last result to form the equation-of-motion for the Green's function in (5.22) we find that it satisfies,

$$\omega\langle\langle p_\alpha(l); u_\beta(l')\rangle\rangle = \frac{i}{2\pi}\delta_{\alpha\beta}\delta_{ll'}$$

$$+ i \sum_{l''\gamma} Y_{\alpha\gamma}(l, l'')P_{\gamma\beta}(l'', l'; \omega). \tag{5.24}$$

ISBN 0-8053-6610-5, 0-8053-6611-3 (pbk.)

Hence,

$$M\omega^2 P_{\alpha\beta}(l, l'; \omega) = \frac{1}{2\pi}\delta_{\alpha\beta}\delta_{ll'} + \sum_{l''\gamma} Y_{\alpha\gamma}(l, l'')P_{\gamma\beta}(l'', l'; \omega) \tag{5.25}$$

or,

$$\sum_{l''\gamma} \{M\omega^2\delta_{\alpha\gamma}\delta_{ll''} - Y_{\alpha\gamma}(l, l'')\}P_{\gamma\beta}(l'', l'; \omega) = \frac{1}{2\pi}\delta_{\alpha\beta}\delta_{ll'}. \tag{5.26}$$

To solve this equation we expand $P_{\alpha\beta}(l, l'; \omega)$ in terms of the eigenvectors of the force-constant matrix. The eigenvectors $\boldsymbol{\sigma}^j(\mathbf{q})$ are real and even in $\mathbf{q}$ for a Bravais lattice, and j labels the three phonon branches and the density of vectors $\mathbf{q}$ is, for each value of j, $\Omega/(2\pi)^3$, where Ω is the volume of the crystal. The phonon frequencies $\omega_j(\mathbf{q})$ are defined by the eigenvalue equation

$$\sum_{l''\gamma} Y_{\alpha\gamma}(l, l'')\exp(i\mathbf{q}\cdot\mathbf{l}'')\sigma^j_\gamma(\mathbf{q}) = M\omega_j^2(\mathbf{q})\exp(i\mathbf{q}\cdot\mathbf{l})\sigma^j_\beta(\mathbf{q}) \tag{5.27}$$

and the eigenvectors are normalized such that

$$\sum_j \sigma^j_\alpha(\mathbf{q})\sigma^j_\beta(\mathbf{q}) = \delta_{\alpha\beta} \tag{5.28}$$

and

$$\sum_\alpha \sigma^j_\alpha(\mathbf{q})\sigma^{j'}_\alpha(\mathbf{q}) = \delta_{jj'}. \tag{5.29}$$

Now, from (5.27) it follows that

$$\sum_{l''\gamma} \{M\omega^2\delta_{\alpha\gamma}\delta_{ll''} - Y_{\alpha\gamma}(l, l'')\}\exp(i\mathbf{q}\cdot\mathbf{l}'')\sigma^j_\gamma(\mathbf{q}) = M\{\omega^2 - \omega_j^2(\mathbf{q})\}\exp(i\mathbf{q}\cdot\mathbf{l})\sigma^j_\alpha(\mathbf{q}) \tag{5.30}$$

so that, if

$$P_{\alpha\beta}(l, l'; \omega) = \frac{1}{NM}\sum_{j\mathbf{q}} \sigma^j_\alpha(\mathbf{q})\sigma^j_\beta(\mathbf{q})\exp\{i\mathbf{q}\cdot(\mathbf{l} - \mathbf{l}')\}\mathscr{P}^j(\mathbf{q}, \omega) \tag{5.31}$$

where N is the number of atoms in the crystal (= number of unit cells) then, from (5.26) $\mathscr{P}^j(\mathbf{q}, \omega)$ satisfies

$$\frac{2\pi}{N}\sum_{j\mathbf{q}} \{\omega^2 - \omega_j^2(\mathbf{q})\}\sigma^j_\alpha(\mathbf{q})\sigma^j_\beta(\mathbf{q})\exp\{i\mathbf{q}\cdot(\mathbf{l} - \mathbf{l}')\}\mathscr{P}^j(\mathbf{q}, \omega) = \delta_{\alpha\beta}\delta_{ll'}. \tag{5.32}$$

From the orthogonality condition (5.28) and

$$\sum_{\mathbf{q}} \exp\{i\mathbf{q}\cdot(\mathbf{l} - \mathbf{l}')\} = N\delta_{ll'} \tag{5.33}$$

ISBN 0-8053-6610-5, 0-8053-6611-3 (pbk.)

it then follows that

$$\mathscr{P}^j(\mathbf{q}, \omega) = (2\pi\{\omega^2 - \omega_j^2(\mathbf{q})\})^{-1}. \tag{5.34}$$

The result (5.34) has exactly the same form as the Green's function for a harmonic oscillator (1.80) if we identify the characteristic frequency of the oscillator with the phonon frequency.

The displacement autocorrelation function is obtained from the general formula (5.11), which for the present problem leads to

$$\langle u_\beta(l')u_\alpha(l, t)\rangle = 2\int_{-\infty}^{\infty} d\omega \exp(i\omega t)$$
$$\{1 - \exp(-\beta\omega)\}^{-1}P''_{\alpha\beta}(l, l'; \omega). \tag{5.35}$$

The imaginary part of the Green's function in (5.35) is evaluated from (5.31) and (5.34) with the substitution $\omega \to \omega - i\eta$ and $\eta \to 0$. We find the result

$$\langle u_\beta(l')u_\alpha(l, t)\rangle$$
$$= \int_{-\infty}^{\infty} d\omega \exp(i\omega t)\{1 - \exp(-\beta\omega)\}^{-1}\frac{1}{NM}\sum_{j\mathbf{q}} \sigma_\alpha^j(\mathbf{q})\sigma_\beta^j(\mathbf{q})$$
$$\times \exp\{i\mathbf{q}\cdot(\mathbf{l} - \mathbf{l}')\}\frac{1}{2\omega_j(\mathbf{q})}[\delta(\omega - \omega_j) - \delta(\omega + \omega_j)]$$
$$= \frac{1}{NM}\sum_{j\mathbf{q}} \sigma_\alpha^j(\mathbf{q})\sigma_\beta^j(\mathbf{q})$$
$$\times \exp\{i\mathbf{q}\cdot(\mathbf{l} - \mathbf{l}')\}\frac{1}{2\omega_j(\mathbf{q})}[\{1 + n_j(\mathbf{q})\}\exp(i\omega_j(\mathbf{q})t)$$
$$+ n_j(\mathbf{q})\exp(-i\omega_j(\mathbf{q})t)], \tag{5.36}$$

where the phonon occupation function

$$n_j(\mathbf{q}) = (\exp\{\beta\omega_j(\mathbf{q})\} - 1)^{-1}. \tag{5.37}$$

Let us now calculate the Green's function for the mass-impurity problem. Using the Hamiltonian (5.17) it is not difficult to show that the Green's function satisfies

$$\sum_{l''\gamma} \{M\omega^2\delta_{\alpha\gamma}\delta_{ll''} - Y_{\alpha\gamma}(l, l'')\}G_{\gamma\beta}(l'', l'; \omega)$$
$$= \frac{1}{2\pi}\delta_{\alpha\beta}\delta_{ll'} + \sum_{l''\gamma} V_{\alpha\gamma}(l, l'')G_{\gamma\beta}(l'', l'; \omega), \tag{5.38}$$

where the perturbation matrix $\mathbf{V}$ is defined by,

$$V_{\alpha\beta}(l, l') = M\lambda\omega^2\sum_e \delta_{\alpha\beta}\delta_{le}\delta_{el'}. \tag{5.39}$$

ISBN 0-8053-6610-5, 0-8053-6611-3 (pbk.)

We shall now write (5.38) in terms of the unperturbed Green's function P. For this it is convenient to employ a matrix notation and write (5.26) as

$$(M\omega^2\mathbf{I} - \mathbf{Y})\cdot\mathbf{P} = \frac{1}{2\pi}\mathbf{I} \tag{5.40}$$

where **I** is the unit matrix. The equation-of-motion for the perturbed Green's function (5.38) in the new notation is

$$(M\omega^2\mathbf{I} - \mathbf{Y})\cdot\mathbf{G} = \frac{1}{2\pi}\mathbf{I} + \mathbf{V}\cdot\mathbf{G}. \tag{5.41}$$

From (5.40),

$$(M\omega^2\mathbf{I} - \mathbf{Y}) = \frac{1}{2\pi}\mathbf{P}^{-1}$$

and using this result in (5.41)

$$\frac{1}{2\pi}\mathbf{P}^{-1}\cdot\mathbf{G} = \frac{1}{2\pi}\mathbf{I} + \mathbf{V}\cdot\mathbf{G}$$

or

$$\mathbf{G} = \mathbf{P} + 2\pi\mathbf{P}\cdot\mathbf{V}\cdot\mathbf{G} \tag{5.42}$$

which is the desired result. Equation (5.42) is usually referred to as the Dyson equation for the Green's function **G**. An elementary survey of the properties of Dyson's equation is given by Ziman (1969), for example.

From (5.42) we have immediately,

$$\mathbf{G} = (\mathbf{I} - 2\pi\mathbf{P}\cdot\mathbf{V})^{-1}\cdot\mathbf{P} \tag{5.43}$$

so that the poles of the perturbed Green's function are determined by the condition,

$$\text{Re Det}|\mathbf{I} - 2\pi\mathbf{P}\cdot\mathbf{V}| = 0. \tag{5.44}$$

Quantities of physical interest are obtained from the Green's function (5.43) averaged over the distribution of impurities. Such configurational averages are difficult to perform well, and we shall give only an elementary discussion at the end of this section. The reader interested in this problem is referred to reviews by Cowley and Buyers (1972) and Elliott *et al.* (1974) for additional details and references to original papers.

The case of a single impurity can be solved exactly, and we shall now obtain the Green's functions and discuss the effect of the impurity on the vibrations of the host atoms.

Let the single impurity be located at the site $e = 0$. The Dyson equation (5.42) then reduces to

$$G_{\alpha\beta}(l, l'; \omega) = P_{\alpha\beta}(l, l'; \omega) + 2\pi\lambda\omega^2 M \sum_{\gamma} P_{\alpha\gamma}(l, 0; \omega) G_{\gamma\beta}(0, l'; \omega). \tag{5.45}$$

ISBN 0-8053-6610-5, 0-8053-6611-3 (pbk.)

For cubic crystals, to which we now restrict our attention, the unperturbed Green's function $P_{\alpha\beta}(l, l; \omega)$ is independent of the Cartesian labels. We therefore define for these crystals

$$P(\omega) = 2\pi M P_{\alpha\alpha}(l, l; \omega)$$
$$= \frac{1}{3N} \sum_{j\mathbf{q}} \{\omega^2 - \omega_j^2(\mathbf{q})\}^{-1}. \quad (5.46)$$

The factor $\frac{1}{3}$ in (5.46) arises since, for all α, $|\sigma_\alpha^j|^2 = \frac{1}{3}$ by virtue of (5.28).

Setting $l = 0$ in (5.45) and using (5.46) we find

$$G_{\alpha\beta}(0, l'; \omega) = \{1 - \lambda\omega^2 P(\omega)\}^{-1} P_{\alpha\beta}(0, l'; \omega) \quad (5.47)$$

and substituting this result into (5.45) we obtain the complete solution for the perturbed Green's function,

$$G_{\alpha\beta}(l, l'; \omega) = P_{\alpha\beta}(l, l'; \omega) + 2\pi\lambda\omega^2 M\{1 - \lambda\omega^2 P(\omega)\}^{-1}$$
$$\times \sum_{\gamma} P_{\alpha\gamma}(l, 0; \omega) P_{\gamma\beta}(0, l'; \omega). \quad (5.48)$$

The second term in (5.48) is the contribution to the Green's function created by the scattering of phonons from the mass-defect; the scattered amplitude is proportional to an unperturbed Green's function, and the scattering cross section is evidently very dependent on the frequency.

From these results it follows that new features in the spectrum of lattice vibrations due to the impurity are determined by the solutions of the equation,

$$\mathrm{Re}\{1 - \lambda\omega^2 P(\omega)\} = 0. \quad (5.49)$$

To proceed in the study of this equation we require a knowledge of the function $P(\omega)$ defined in (5.46). A calculation of $P(\omega)$ for a realistic model of a crystal requires a model for the phonon spectrum, and then the numerical integration over the Brillouin zone of $\{\omega^2 - \omega_j^2(\mathbf{q})\}^{-1}$ for a range of values of ω. However, we should obtain the salient features by using a Debye model for the lattice vibrations, and for which $P(\omega)$ can be obtained in analytic form. To this end we write $P(\omega)$ in terms of the normalized phonon density of states* $Z(\omega)$,

$$P(\omega) = \lim_{\eta \to 0} \int_0^{\omega_D} \frac{du\, Z(u)}{(\omega - i\eta)^2 - u^2}$$
$$= P\int_0^{\omega_D} \frac{du\, Z(u)}{\omega^2 - u^2} + \frac{i\pi}{2\omega} Z(\omega) \quad (5.50)$$

where ω_D is the Debye frequency and P denotes the principal part of the

* The density of states can be measured by inelastic, incoherent neutron scattering, cf. Squires (1978).

ISBN 0-8053-6610-5, 0-8053-6611-3 (pbk.)

integral. For a Debye model of a solid

$$Z(\omega) = 3\omega^2/\omega_D^3, \qquad \omega \leq \omega_D$$
$$= 0, \qquad \omega > \omega_D$$

and carrying out the integration in (5.50), Eq. (5.49) reduces to

$$\lambda^{-1} = y(x) = -3x^2 + \tfrac{3}{2}x^3 \ln|(1 + x)/(1 - x)|, \tag{5.51}$$

where the reduced frequency $x = (\omega/\omega_D)$. The function $y(x)$ is shown in Fig. (5.1) for $0 \leq x \leq 3$. The singular behavior at $\omega = \omega_D$ is an artifact of the simple model used for the phonon density of states, and for a more realistic model $P(\omega)$ is finite at the edge of the phonon frequency band.

Because λ^{-1} can only have values in the range ∞ to 1, and 0 to $-\infty$, there is a range $0 \leq \lambda^{-1} \leq 1$ in which there are no solutions of (5.51). For light mass-impurities $M' < M$ two types of solution can exist for which either $\omega < \omega_D$ or $\omega > \omega_D$. Because the density of states $Z(\omega)$ is zero for $\omega > \omega_D$ it follows that solutions of (5.51) for $\omega > \omega_D$ correspond to impurity modes with an infinite lifetime. On the other hand solutions that occur for $\omega < \omega_D$ correspond to impurity modes that are damped by interaction with the phonons. The damping of these modes is proportional to $Z(\omega)$ and hence

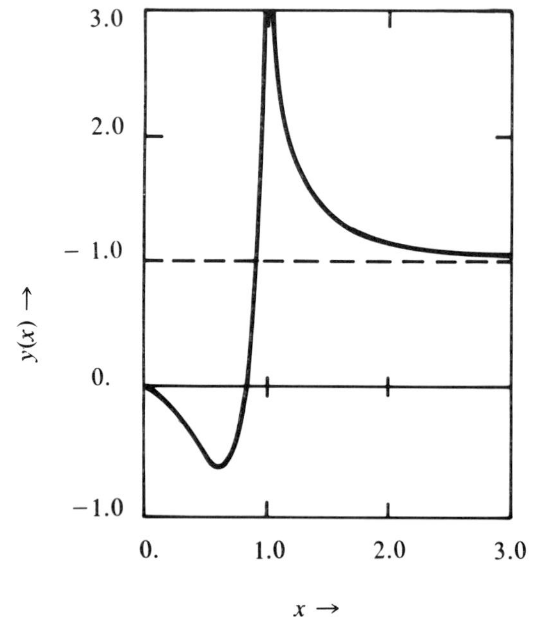

Fig. (5.1) The function $y(x)$ defined by Eq. (5.51) is shown for x in the range $0 \leq x \leq 3$.

ISBN 0-8053-6610-5, 0-8053-6611-3 (pbk.)

those near the band edge will not strongly influence the properties of the crystal. The in-band light mass-impurity modes only occur near the band edge, and they are therefore of minimal importance in determining the properties of the mixed crystal. Also the modes with $\omega > \omega_D$ will not influence the thermodynamic properties of the crystal because of their high energy, although they can be observed in neutron scattering experiments. Heavy mass-impurity modes occur in pairs for (M'/M) greater than some critical value, which is approximately 2.6 for the Debye model. The lowest frequency heavy mass-impurity mode can have a very pronounced effect on the thermodynamic properties of the mixed crystal since it can be highly populated at modest temperatures and its damping, through interactions with the phonons, is small, because $Z(\omega)$ is small at low frequencies; cf. Elliott *et al.* (1974).

We shall now calculate the contribution to the displacement autocorrelation function made by a light mass-impurity mode that occurs above the band edge. This correlation function determines the contribution made by the mode to the cross section for radiation scattering [Elliott *et al.* (1974)] and it is proportional to the imaginary part of the Green's function (5.48),

$$G''_{\alpha\beta}(l, l'; \omega) = 2\pi\lambda\omega^2 M \sum_{\gamma} P_{\alpha\gamma}(l, 0; \omega)P_{\gamma\beta}(0, l'; \omega) \times \text{Im}\{1 - \lambda\omega^2 P(\omega)\}^{-1}. \tag{5.52}$$

In obtaining this result we have used the fact that the imaginary part of the unperturbed Green's function is zero when ω is above the band edge. If the function $g(z)$ satisfies $g^*(z) = g(z^*)$, and $z = \omega - i\eta$, and $g(\omega_0) = 0$, then using the Taylor expansion, $g(z) \simeq (\omega - i\eta - \omega_0)(dg/d\omega)_{\omega_0}$, we find,

$$\begin{aligned}\lim_{\eta\to 0} \text{Im}\, g^{-1}(z) &= \frac{1}{2i}\lim_{\eta\to 0}\{g^{-1}(z) - g^{-1}(z^*)\} \\ &= \frac{1}{2i}\left(\frac{dg}{d\omega}\right)^{-1}_{\omega_0} \lim_{\eta\to 0}\left\{\frac{1}{\omega - i\eta - \omega_0} - \frac{1}{\omega + i\eta - \omega_0}\right\} \\ &= \pi\left(\frac{dg}{d\omega}\right)^{-1}_{\omega_0} \delta(\omega - \omega_0).\end{aligned} \tag{5.53}$$

For the present case, $g = 1 - \lambda\omega^2 P(\omega)$,

$$\frac{d}{d\omega}\{1 - \lambda\omega^2 P(\omega)\} = 2\lambda\omega \int_0^{\omega_D} \frac{du\, u^2\, Z(u)}{(\omega^2 - u^2)^2}, \qquad \omega > \omega_D$$

and the integral is positive. The displacement Green's function is, for

ISBN 0-8053-6610-5, 0-8053-6611-3 (pbk.)

$\omega > \omega_D$,

$$G''_{\alpha\beta}(l, l'; \omega) = \pi^2 \omega_0 M\{\delta(\omega - \omega_0) - \delta(\omega + \omega_0)\}$$
$$\times \left(\int_0^{\omega_D} \frac{du\, u^2\, Z(u)}{(\omega_0^2 - u^2)^2} \right)^{-1}$$
$$\times \sum_\gamma P_{\alpha\gamma}(l, 0; \omega_0) P_{\gamma\beta}(0, l'; \omega_0) \tag{5.54}$$

where we have taken account of the fact that there are two solutions of $g(\omega) = 0$, namely $\omega = \pm\omega_0$. Notice that there is no explicit dependence in (5.54) on the perturbation parameter λ; the dependence on λ, for a given phonon spectrum, is contained in the value of the impurity mode frequency, ω_0.

A displacement autocorrelation function of particular interest is the one for the impurity atom, and this is obtained from

$$G''_{\alpha\beta}(0, 0; \omega) = \delta_{\alpha\beta} \left(\frac{\omega_0}{4M} \right) P^2(\omega_0) \left(\int_0^{\omega_D} \frac{du\, u^2\, Z(u)}{(\omega_0^2 - u^2)^2} \right)^{-1}$$
$$\times \{\delta(\omega - \omega_0) - \delta(\omega + \omega_0)\}. \tag{5.55}$$

In deriving this result from (5.52) we have used the fact that $P_{\alpha\beta}(l, l; \omega)$ is diagonal in the Cartesian component labels, and (5.46). If $\omega_0 \gg \omega_D$ then $P(\omega_0)$ can be evaluated as a power series in (ω_D/ω_0),

$$P(\omega_0) = \omega_0^{-2} \int_0^{\omega_D} du\, Z(u)[1 + (u/\omega_0)^2 + \cdots]$$
$$= \omega_0^{-2}[1 + \tfrac{3}{5}(\omega_D/\omega_0)^2 + \cdots].$$

Similarly,

$$\int_0^{\omega_D} \frac{du\, u^2\, Z(u)}{(\omega_0^2 - u^2)^2} = \frac{3}{5} \frac{\omega_D^2}{\omega_0^4} \left[1 + \frac{10}{7} (\omega_D/\omega_0)^2 + \cdots \right].$$

From these results we find, for $\omega_0 \gg \omega_D$,

$$G''_{\alpha\beta}(0, 0; \omega) \simeq \delta_{\alpha\beta} \left(\frac{5\omega_0}{12M\omega_D^2} \right) \left[1 - \frac{8}{35} (\omega_D/\omega_0)^2 + \cdots \right]$$
$$\times \{\delta(\omega - \omega_0) - \delta(\omega + \omega_0)\}, \tag{5.56}$$

and the corresponding displacement autocorrelation function is, from (5.35),

$$\langle u_\beta(l' = 0) u_\alpha(l = 0, t) \rangle$$
$$= \delta_{\alpha\beta} \left(\frac{5\omega_0}{6M\omega_D^2} \right) \left[1 - \frac{8}{35} (\omega_D/\omega_0)^2 + \cdots \right]$$
$$\times \{(1 + n_0) \exp(i\omega_0 t) + n_0 \exp(-i\omega_0 t)\} \tag{5.57}$$

ISBN 0-8053-6610-5, 0-8053-6611-3 (pbk.)

where the occupation factor,

$$n_0 = (\exp\{\beta\omega_0\} - 1)^{-1}$$

is very small if $\omega_0 \gg k_B T$. We conclude from this result for the impurity displacement autocorrelation function that the displacement is large for a light mass-impurity for which the associated impurity mode frequency is well above the phonon band edge. In consequence, the impurity is a strong scatterer of incident radiation that couples to the displacement, e.g. thermal neutrons. Wakabayashi *et al.* (1971) have compared theory and neutron scattering measurements for Ge doped with a low concentration of the lighter mass element Si, and they find that the theory gives a very good account of the observed impurity mode.

We close this section by considering the calculation of the displacement Green's function for a small concentration c of mass-impurities. Let us also assume that the defect on the host lattice about each impurity is strongly localized so that the cluster formed by an impurity and its associated defect can be regarded as an isolated scatterer of lattice waves. The probability of a lattice wave scattering from one cluster is proportional to c, the probability that it suffers scattering by a second cluster is proportional to c^2, and so on and so forth. These scattering events are additive in the displacement Green's function, at least for very small c, as can be seen from the exact solution for a single scatterer, Eq. (5.48). If we expand $G_{\alpha\beta}(l, l'; \omega)$ in terms of the eigenvectors of the force constant matrix, as in (5.31), then we find from (5.48) that the spatial Fourier transform of the exact Green's function is

$$G(\mathbf{q}, \omega) = \mathscr{P}^j(\mathbf{q}, \omega)\left\{1 + \frac{2\pi}{N} T(\omega)\mathscr{P}^j(\mathbf{q}, \omega)\right\} \tag{5.58}$$

where

$$T(\omega) = \lambda\omega^2/\{1 - \lambda\omega^2 P(\omega)\}. \tag{5.59}$$

Because $c = 1/N$ for a single defect, (5.58) shows that a single scattering event adds a term to the Green's function which is proportional to $c\{\mathscr{P}^j(\mathbf{q}, \omega)\}^2$. The Green's function for a finite concentration of defects has full translational invariance after it has been averaged over the configuration of defects, and it can therefore be spatially Fourier transformed in terms of a function $\bar{G}^j(\mathbf{q}, \omega)$, where the bar denotes the configurational average. The foregoing discussion and the form of the Green's function for a single defect (5.58) lead to the result

$$\begin{aligned} \bar{G}^j(\mathbf{q}, \omega) &= \mathscr{P}^j(\mathbf{q}, \omega)\{1 + 2\pi c T(\omega)\mathscr{P}^j(\mathbf{q}, \omega) \\ &\qquad + [2\pi c T(\omega)\mathscr{P}^j(\mathbf{q}, \omega)]^2 + \cdots\} \\ &= \mathscr{P}^j(\mathbf{q}, \omega)\{1 - 2\pi c T(\omega)\mathscr{P}^j(\mathbf{q}, \omega)\}^{-1} \\ &= (2\pi\{\omega^2 - \omega_j^2(\mathbf{q}) - cT(\omega)\})^{-1}. \end{aligned} \tag{5.60}$$

ISBN 0-8053-6610-5, 0-8053-6611-3 (pbk.)

The dispersion of the collective modes of the mixed system are now determined by solutions of the equation

$$\omega_0^2 = \omega_j^2(\mathbf{q}) + c \,\mathrm{Re}\, T(\omega_0). \tag{5.61}$$

If the width of the mode $\Gamma(\mathbf{q}, \omega_0)$ is small compared to $\omega_j(\mathbf{q})$, then to a good approximation

$$\Gamma(\mathbf{q}, \omega_0) = \frac{\pi c \lambda^2 \omega_0^3 Z(\omega_0)/4\omega_j(\mathbf{q})}{\{1 - \lambda\omega_0^2 P'(\omega_0)\}^2 + \{\lambda\omega_0^2 P''(\omega_0)\}^2}. \tag{5.62}$$

Here we have written $P(\omega) = P'(\omega) + iP''(\omega)$ and expressed $P''(\omega)$ in the numerator in terms of the density of states using (5.50). Notice that the width induced by the impurities (5.62) is proportional to the unperturbed phonon density of states and the impurity concentration, and that it is explicitly of second order in the perturbation parameter λ.

The results we have derived are very similar to those obtained with an approximation scheme called the average t-matrix approximation. In this approximation, which is reviewed in detail by Elliott *et al.* (1974), there is an additional factor of $(1 - c)$ that multiplies $P(\omega)$ in the denominator of $T(\omega)$, which is then the average t-matrix of the theory. This additional factor represents only a small correction for small values of c. The predictions of the average t-matrix approximation have been compared with results obtained by neutron scattering experiments and it is found to be reasonably satisfactory for simple mixed systems and $c \lesssim 10\%$. A notable defect is a lack of structure in the calculated widths (5.62), and this seems to be attributable to a dependence of the force constants on the masses rather than a failure of the average t-matrix approximation.

5.3 ELECTRON GAS

The importance of studying the properties of an electron gas hardly needs emphasizing, and an introduction to the subject can be found in any text on solid state physics. Because of its importance in the quantum theory of solids the electron gas, or plasma, has been studied by many authors using a wide variety of computational techniques. Many of the important results were derived first using diagrammatic perturbation theory, and the aim of the present section is to illustrate how some of these results can be derived with causal Green's functions. The properties of an electron gas subject to a steady magnetic field are discussed in Chap. 6, using linear response theory to derive appropriate expressions, and formulae derived there reduce for zero field to the corresponding results obtained in this section.

The simplest, nontrivial approximation to the correlation effects is the Hartree-Fock theory in which correlations are due solely to the Pauli exclusion principle, i.e. effects due to the Coulomb interaction are neglected. In the Hartree-Fock theory the energy of an electron is the sum of the free

ISBN 0-8053-6610-5, 0-8053-6611-3 (pbk.)

electron kinetic energy ($k^2/2m$), and a term which has an infinite slope as the wavevector, k, approaches the Fermi surface. The latter feature has an interesting and catastrophic consequence since it makes the electron density of states zero at the Fermi surface, whereas many experiments on solids indicate that the electron density of states is finite at the Fermi surface. The origin of this catastrophe with Hartree-Fock theory can be traced back to the long-range nature of the Coulomb interaction which has a spatial Fourier transform that diverges in the limit of small wavevectors.

If we calculate the internal energy, which at $T = 0$ (a degenerate electron gas) is the same as the free-energy, the Hartree-Fock theory gives a sensible result, namely, the average kinetic energy plus the (negative) exchange energy, cf. Eq. (5.89). The reason why the Hartree-Fock theory is reasonable for the internal energy is that only a small proportion of electrons with wavevectors close to the Fermi surface are effected by the anomalous behavior of the Coulomb interaction, and the internal energy involves an integral over all wavevectors. The calculation of the internal energy is particularly important since, for a range of simple materials, the difference between the Hartree-Fock value, Eq. (5.89), and the energy obtained with a more sophisticated theory contains practically all the measured cohesive energy.

Attempts to go beyond Hartree-Fock theory by expanding in the Coulomb potential give nonsensical results if the expansions are terminated at a low order. Here, again, the root cause of the trouble is the long-range nature of the Coulomb interaction. The calculation of corrections to Hartree-Fock theory actually requires the summation of infinite numbers of terms in the perturbation expansion, and these summations can be readily accomplished using Green's functions and decoupling procedures.

Electrons in a solid are subject to a Coulomb interaction and a periodic potential due to ion cores. For simple metals, like sodium, the main effect of the periodic potential is accounted for by introducing an effective electron mass $m^* \neq m_e$. Because the Block wavefunctions are also plane waves, to a good approximation the free electron gas is a good model for the conduction band electrons in simple metals. The same is true for the elements of groups I, II, III and IV of the periodic table, all of which have fairly free-electron-like band structures.

Probably the most useful function for describing the properties of an electron gas is the wavevector and frequency dependent dielectric function, $\epsilon(\mathbf{k}, \omega)$. From a knowledge of this function we can obtain, for example, the dispersion of collective modes in the gas, and the total energy.

The dielectric function can be defined by considering the change in the potential energy of an electron gas due to a wavevector and frequency dependent test charge perturbation. In this instance the dielectric function is given by the applied potential divided by the total potential, which is the sum of the applied and induced potentials, cf. Pines and Nozières (1966).

ISBN 0-8053-6610-5, 0-8053-6611-3 (pbk.)

We choose to define the dielectric function in terms of the Green's function for the electron density whose **k**th Fourier component is

$$n_{\mathbf{k}} = \sum_j \exp(-i\mathbf{k}\cdot\mathbf{R}_j) \tag{5.63}$$

where the sum is over all N electrons in a volume Ω. Since there is assumed to be a uniform positive charge background of density $n = N/\Omega$ the total charge of the system is zero. The dielectric function is defined by the equation

$$\epsilon^{-1}(\mathbf{k}, \omega) = 1 + \left(\frac{4\pi e^2}{k^2\Omega}\right) 2\pi\langle\langle n_{\mathbf{k}}; n_{\mathbf{k}}^{+}\rangle\rangle \tag{5.64}$$

where e is the magnitude of the charge on an electron. Notice that $\epsilon(\mathbf{k}, \omega)$ is dimensionless, and that the dispersion of the excitations in the gas are determined by the solutions of the equation $\epsilon(\mathbf{k}, \omega) = 0$.

The Hamiltonian of an interacting homogeneous electron gas is

$$\mathcal{H} = \sum_j \frac{1}{2m^*}\mathbf{p}_j^2 + \tfrac{1}{2}\sum_{\mathbf{k}}{}' \left(\frac{4\pi e^2}{k^2\Omega}\right)(n_{\mathbf{k}}^{+} n_{\mathbf{k}} - N). \tag{5.65}$$

Here m^* is the effective mass of an electron, and the prime on the summation in the second term indicates that the term with $\mathbf{k} = 0$ is omitted, since it cancels against the uniform background of positive charge. The total energy can be obtained from the dielectric function by integrating it with respect to the coupling constant e^2 in (5.65). The first step in the derivation of this result is to note that the interaction energy can be simply expressed in terms of the dielectric function. From (5.65) it follows that the interaction energy

$$\begin{aligned} E_{\text{int}} &= \tfrac{1}{2}\sum_{\mathbf{k}}{}' \left(\frac{4\pi e^2}{k^2\Omega}\right)\langle\{n_{\mathbf{k}}^{+} n_{\mathbf{k}} - N\}\rangle \\ &= \tfrac{1}{2}\sum_{\mathbf{k}}{}' \left(\frac{4\pi e^2 n}{k^2}\right)\{S(k) - 1\} \end{aligned} \tag{5.66}$$

where $S(k)$ is the structure factor

$$S(k) = \frac{1}{N}\langle n_{\mathbf{k}}^{+} n_{\mathbf{k}}\rangle = \int_{-\infty}^{\infty} d\omega\, S(k, \omega) \tag{5.67}$$

and

$$S(k, \omega) = \frac{1}{2\pi N}\int_{-\infty}^{\infty} dt\, \exp(-i\omega t)\langle n_{\mathbf{k}}^{+} n_{\mathbf{k}}(t)\rangle. \tag{5.68}$$

ISBN 0-8053-6610-5, 0-8053-6611-3 (pbk.)

From (5.11) and (5.64) we have

$$S(k, \omega) = \frac{2}{N}\{1 - \exp(-\beta\omega)\}^{-1}\langle\langle n_{\mathbf{k}}; n_{\mathbf{k}}^{+}\rangle\rangle''$$

$$= 2\{1 - \exp(-\beta\omega)\}^{-1}\frac{1}{2\pi}\left(\frac{k^2}{4\pi e^2 n}\right)\text{Im}\,\epsilon^{-1}(\mathbf{k}, \omega) \qquad (5.69)$$

so that (5.66) becomes,

$$E_{\text{int}} = -\sum_{\mathbf{k}}{}'\left(\frac{2\pi e^2 n}{k^2}\right) + \frac{1}{2\pi}\sum_{\mathbf{k}}{}'\int_{-\infty}^{\infty} d\omega$$
$$\times\{1 - \exp(-\beta\omega)\}^{-1}\,\text{Im}\,\epsilon^{-1}(\mathbf{k}, \omega). \qquad (5.70)$$

The total energy is obtained from (5.70) by treating E_{int} as a function of the coupling constant $\lambda = e^2$. For a *degenerate* electron gas the total energy per electron is [Pines and Nozières (1966)],

$$\tfrac{3}{5}E_f + \frac{1}{N}\int_0^1 d(\ln\lambda)E_{\text{int}}(\lambda) \qquad (5.71)$$

where the Fermi energy,*

$$E_f = p_f^2/2m^*, \quad \text{with} \quad p_f = (3\pi^2 n)^{1/3}. \qquad (5.72)$$

The relation (5.67) can be regarded as the zeroth sum-rule for $S(k, \omega)$. The next sum-rule leads to a result that is independent of the Coulomb interaction and temperature. From (5.68) and the equation-of-motion for $n_{\mathbf{k}}$ we have

$$N\int_{-\infty}^{\infty} d\omega\,\omega\,S(k, \omega) = -\langle n_{\mathbf{k}}^{+}[n_{\mathbf{k}}, \mathcal{H}]\rangle$$

$$= -\tfrac{1}{2}\langle[n_{\mathbf{k}}^{+}, [n_{\mathbf{k}}, \mathcal{H}]]\rangle \qquad (5.73)$$

and since $[n_{\mathbf{k}}^{+}, [n_{\mathbf{k}}, \mathcal{H}]] = -Nk^2/m^*$ the sum-rule is

$$\int_{-\infty}^{\infty} d\omega\,\omega\,S(k, \omega) = k^2/2m^*. \qquad (5.74)$$

Additional sum-rules are derived by Pines and Nozières (1966), for example.

After these few preliminary remarks we turn to the calculation of the Green's function $\langle\langle n_{\mathbf{k}}; n_{\mathbf{k}}^{+}\rangle\rangle$, and thereby the dielectric function. For this it is convenient to introduce Fermi creation and annihilation operators c^{+}

* A useful value is $(1/2m_e) = 3.83$ eVÅ^2, where m_e is the mass of an electron.

ISBN 0-8053-6610-5, 0-8053-6611-3 (pbk.)

and c, respectively. These operators satisfy the anticommutation relation

$$[c_{\mathrm{p}\sigma},\, c^{+}_{\mathrm{g}\sigma'}]_{+} = \delta_{\mathrm{pq}}\delta_{\sigma\sigma'},$$
$$[c_{\mathrm{p}\sigma},\, c_{\mathrm{q}\sigma'}]_{+} = [c^{+}_{\mathrm{p}\sigma},\, c^{+}_{\mathrm{q}\sigma'}]_{+} = 0 \tag{5.75}$$

where σ labels the two spin states of an electron. In terms of these operators,

$$n_{\mathrm{k}} = \sum_{\mathrm{p}\sigma} c^{+}_{\mathrm{p}\sigma} c_{\mathrm{k+p}\sigma} \tag{5.76}$$

and the Hamiltonian (we drop the constant term in (5.65) since it will not contribute to an equation-of-motion)

$$\mathscr{H} = \sum_{\mathrm{p}\sigma} E_{p} c^{+}_{\mathrm{p}\sigma} c_{\mathrm{p}\sigma} + \tfrac{1}{2} \sum_{\mathrm{pqk}}{}' \sum_{\sigma\sigma'} \left(\frac{4\pi e^{2}}{k^{2}\Omega}\right) c^{+}_{\mathrm{k+p}\sigma} c^{+}_{\mathrm{q-k}\sigma'} c_{\mathrm{q}\sigma'} c_{\mathrm{p}\sigma} \tag{5.77}$$

with

$$E_{k} = k^{2}/2m^{*}. \tag{5.78}$$

The equation-of-motion for the Green's function of interest contains the commutator, $[c^{+}_{\mathrm{p}\sigma} c_{\mathrm{k+p}\sigma},\, \mathscr{H}]$. The evaluation of this entails calculating commutators of the form $[AB, CD]$ where A, B, C and D are Fermi operators, and for such commutators we have the useful result,

$$[AB, CD] = -[A, C]_{+}BD - C[A, D]_{+}B + A[B, C]_{+}D + CA[B, D]_{+}. \tag{5.79}$$

Using this result together with (5.77),

$$[c^{+}_{\mathrm{p}\sigma} c_{\mathrm{k+p}\sigma},\, \mathscr{H}] = (E_{k+p} - E_{p}) c^{+}_{\mathrm{p}\sigma} c_{\mathrm{k+p}\sigma} + \sum_{\mathrm{qq}'\sigma'}{}' \left(\frac{4\pi e^{2}}{q^{2}\Omega}\right) \{c^{+}_{\mathrm{p}\sigma} c^{+}_{\mathrm{q+q}'\sigma'} c_{\mathrm{q}'\sigma'} c_{\mathrm{k+p+q}\sigma} - c^{+}_{\mathrm{p+q}\sigma} c^{+}_{\mathrm{q}'-\mathrm{q}\sigma'} c_{\mathrm{q}'\sigma'} c_{\mathrm{k+p}\sigma}\}. \tag{5.80}$$

If we neglect the Coulomb interaction for the moment, the equation-of-motion is

$$-\omega\langle\langle c^{+}_{\mathrm{p}\sigma} c_{\mathrm{k+p}\sigma};\, n^{+}_{\mathrm{k}}\rangle\rangle = \frac{1}{2\pi}\langle[c^{+}_{\mathrm{p}\sigma} c_{\mathrm{k+p}\sigma},\, n^{+}_{\mathrm{k}}]\rangle + (E_{k+p} - E_{p})\langle\langle c^{+}_{\mathrm{p}\sigma} c_{\mathrm{k+p}\sigma};\, n^{+}_{\mathrm{k}}\rangle\rangle. \tag{5.81}$$

From this result we obtain the noninteracting Green's function

$$\langle\langle n_{\mathrm{k}};\, n^{+}_{\mathrm{k}}\rangle\rangle = (1/2\pi)P_{\mathrm{k}}(\omega)$$

with

$$P_{\mathrm{k}}(\omega) = \sum_{\mathrm{p}\sigma} (f_{p} - f_{k+p})/(-\omega + E_{p} - E_{k+p}) \tag{5.82}$$

ISBN 0-8053-6610-5, 0-8053-6611-3 (pbk.)

where the Fermi distribution function

$$f_p = \langle c^+_{p\sigma} c_{p\sigma} \rangle = (\exp\{\beta(E_p - \mu)\} + 1)^{-1}, \tag{5.83}$$

is independent of the spin of the electron, and μ is the chemical potential.

For a degenerate gas ($T \to 0$ and $\mu = E_f$) the dissipative part of the noninteracting Green's function can be evaluated in closed form. Setting $\omega \to \omega - i\eta$ and $\eta \to 0$ in (5.82) gives

$$\langle\langle n_{\mathrm{k}}; n^+_{\mathrm{k}} \rangle\rangle'' = \frac{1}{2\pi} \frac{\Omega}{(2\pi)^3} \int d\mathbf{p}\, (f_p - f_{k+p}) \pi \delta(\omega - E_p + E_{k+p})$$

and in the limit $T \to 0$ this is nonzero only in the region indicated in Fig. (5.2). In evaluating the expression it is convenient to use reduced energy and wavevector variables,

$$x = \omega / E_f \qquad \text{and} \qquad y = k / p_f. \tag{5.84}$$

The nonzero components of the noninteracting Green's function are found to be: for $0 \le y \le 2$,

$$\langle\langle n_{\mathrm{k}}; n^+_{\mathrm{k}} \rangle\rangle'' = N(3x/16E_f y), \qquad 0 \le x \le 2y - y^2 \tag{5.85}$$

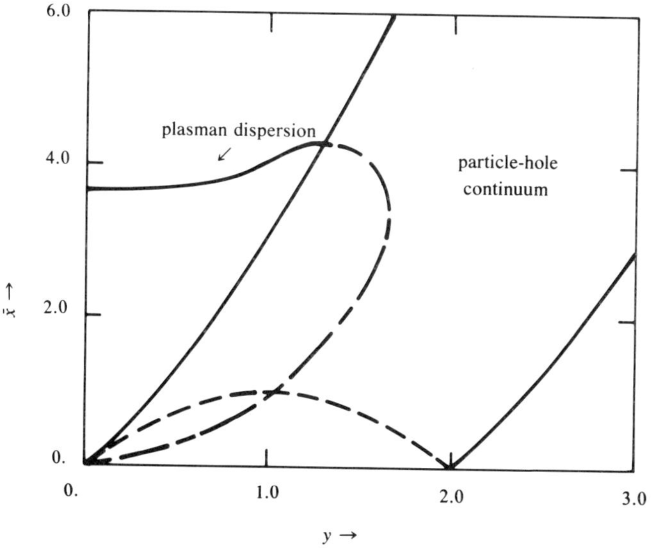

Fig. (5.2) The imaginary part of the noninteracting electron Green's function is nonzero within the particle-hole continuum, bounded by the solid lines, where it has values given by Eq. (5.85); the reduced energy and wavevector units are $x = \omega/E_f$ and $y = k/p_f$. The dispersion of the collective mode which satisfies (5.96) is also sketched.

ISBN 0-8053-6610-5, 0-8053-6611-3 (pbk.)

and

$$\langle\langle n_{\mathbf{k}};\, n_{\mathbf{k}}^{+}\rangle\rangle'' = N(3/16E_f y)\left\{1 - \left(\frac{x - y^2}{2y}\right)^2\right\}; \qquad 2y - y^2 \leq x \leq y^2 + 2y$$

and for $y \geq 2$,

$$\langle\langle n_{\mathbf{k}};\, n_{\mathbf{k}}^{+}\rangle\rangle'' = N(3/16E_f y)\left\{1 - \left(\frac{x - y^2}{2y}\right)^2\right\},$$

$$y^2 - 2y \leq x \leq y^2 + 2y.$$

The corresponding structure factor is ($k = yp_f$)

$$\begin{aligned} S(k) &= \tfrac{3}{4}y(1 - \tfrac{1}{12}y^2), \qquad 0 \leq y \leq 2 \\ &= 1, \qquad 2 \leq y. \end{aligned} \tag{5.86}$$

The Hartree-Fock expression for the ground-state energy is obtained from the general expression (5.71) using (5.86) in (5.66), which merely takes into account the correlations in the electron positions brought about by the Pauli principle. Because E_{int} is linear in e^2 in this approximation the ground-state energy is, from (5.71),

$$\tfrac{3}{5}E_f + \frac{1}{2N}\sum_{\mathbf{k}}{}' \left(\frac{4\pi e^2 n}{k^2}\right)\{S(k) - 1\} \tag{5.87}$$

with $S(k)$ given by (5.86).

It is usual to express the result for electrons ($m^* = m_e$) in units of Rhydbergs, and a dimensionless parameter r_s which is the interelectron spacing measured in units of the Bohr radius a_0. A Rydberg $= m_e e^4/2$,

$$r_s = (3/4\pi a_0^3 n)^{1/3}, \tag{5.88}$$

the kinetic energy of an electron is just k^2 Rhydbergs, and

$$p_f = (9\pi/4)^{1/3} r_s^{-1} = 1.94/r_s.$$

With these units the Hartree-Fock result (5.87) for the ground-state energy per electron is,

$$(2.21/r_s^2) - (0.916/r_s) \qquad \text{(Rhydbergs).} \tag{5.89}$$

The first term is simply the average kinetic energy per electron and the second term is the exchange energy that arises from the correlations induced by the Pauli exclusion principle. We see from this that r_s is a measure of the ratio of the lowest-order potential energy term, the exchange energy, to the average kinetic energy. Therefore, the condition $r_s < 1$, which is achieved with a high density electron gas, corresponds to a weak coupling regime.

Most metals have values of r_s in the range $2 \leq r_s \leq 6$. In consequence,

ISBN 0-8053-6610-5, 0-8053-6611-3 (pbk.)

predictions obtained for a dense, degenerate electron gas are not likely to be quantitatively satisfactory for the interpretation of data for metals. However, the discrepancies between the model and observed properties are smaller than might appear possible on the basis of a discussion using r_s to characterize the systems. The weak coupling condition is well satisfied for a number of doped, degenerate semiconductors for which $m^* \ll m_e$ and the dielectric constant ϵ_0 is very large, since both these trends increase the effective Bohr radius $a_0^* = \epsilon_0/m^*e^2$. For example, for n-GaAs with $n = 10^{17}$ electrons/cm^3, estimates of the effective mass and dielectric constant based on experiments are $m^* \sim 0.08m_e$ and $\epsilon_0 \sim 12$.

We now address the problem of accounting for the interaction term in the commutator (5.80). The terms in the equation-of-motion for $c^+_{\mathbf{p}\sigma}c_{\mathbf{k}+\mathbf{p}\sigma}$ generated by the Coulomb interaction become linear in the operator if we employ a pairing approximation for the products of four operators. For example, if we remember the restriction $\mathbf{q} \neq 0$ in the interaction term, the first product on the right-hand side of (5.80) reduces to the sum of three terms

$$\begin{aligned} c^+_{\mathbf{p}\sigma}c^+_{\mathbf{q}+\mathbf{q}'\sigma'}c_{\mathbf{q}'\sigma'}c_{\mathbf{k}+\mathbf{p}+\mathbf{q}\sigma} = & -\delta_{\sigma\sigma'}\delta_{\mathbf{p}\mathbf{q}'}\, f_p c^+_{\mathbf{p}+\mathbf{q}\sigma}c_{\mathbf{k}+\mathbf{p}+\mathbf{q}\sigma} \\ & + \delta_{\mathbf{k},-\mathbf{q}}\, f_p c^+_{\mathbf{q}'-\mathbf{k}\sigma'}c_{\mathbf{q}'\sigma'} \\ & - \delta_{\sigma\sigma'}\delta_{\mathbf{k}+\mathbf{p},\mathbf{q}'}\, f_{q+q'} c^+_{\mathbf{p}\sigma}c_{\mathbf{q}'\sigma}\,. \end{aligned} \tag{5.90}$$

The relative signs of the terms are determined by the Fermi character of the operators and the number of operators that separate two that are being paired together; if an odd number of operators separates a pair, then the contribution the pair makes to the approximate expression for the product is accompanied by a negative sign, whereas the sign is positive if an even number of operators separate the pair. If the second product of four operators in (5.80) is approximated in a similar way, the commutator reduces to

$$\begin{aligned} [c^+_{\mathbf{p}\sigma}c_{\mathbf{k}+\mathbf{p}\sigma}, \mathscr{H}] \simeq & \, (E_{k+p} - E_p)c^+_{\mathbf{p}\sigma}c_{\mathbf{k}+\mathbf{p}\sigma} \\ & + \left(\frac{4\pi e^2}{k^2\Omega}\right)(f_p - f_{k+p})\sum_{\mathbf{q}\sigma'} c^+_{\mathbf{q}\sigma'}c_{\mathbf{k}+\mathbf{q}\sigma'} \\ & - (f_p - f_{k+p})\sum_{\mathbf{q}}\left(\frac{4\pi e^2}{q^2\Omega}\right)c^+_{\mathbf{p}+\mathbf{q}\sigma}c_{\mathbf{k}+\mathbf{p}+\mathbf{q}\sigma} \\ & - c^+_{\mathbf{p}\sigma}c_{\mathbf{k}+\mathbf{p}\sigma}\sum_{\mathbf{q}}\left(\frac{4\pi e^2}{q^2\Omega}\right)(f_{k+p+q} - f_{p+q}) \end{aligned} \tag{5.91}$$

The three interaction terms on the right-hand side are usually referred to as the direct, exchange and self-energy contributions, respectively. For small k the direct term dominates all other terms, and this is the only one we shall retain in the equation-of-motion for the Green's function. This simpli-

ISBN 0-8053-6610-5, 0-8053-6611-3 (pbk.)

fication for small k occurs because of the long-range nature of the Coulomb potential, and it does not hold in the case of a short-range potential where, in fact, the exchange term is particularly important, e.g. liquid He^3. As an extreme case, we find that for a point interaction potential, whose Fourier transform is a constant, the exchange term cancels half of the direct term in (5.91).

The approximation in which only the direct contribution is retained in (5.91) is usually called the random phase approximation (RPA), or the self-consistent-field approximation. In Sec. 6.1 we shall show that in this approximation the electrons can be regarded as a gas of noninteracting electrons which respond to the average, total potential acting on the gas, which is the sum of the applied and induced potentials. Viewed in this way, corrections to the RPA involve taking account of fluctuations about the average total potential.

Keeping only the direct term in (5.91),

$$\begin{aligned}(-\omega + E_p - E_{k+p})&\langle\langle c^+_{\mathbf{p}\sigma} c_{\mathbf{k}+\mathbf{p}\sigma}; n^+_{\mathbf{k}}\rangle\rangle \\ &= \frac{1}{2\pi}(f_p - f_{k+p}) + \left(\frac{4\pi e^2}{k^2\Omega}\right) \\ &\quad\times (f_p - f_{k+p}) \sum_{\mathbf{q}\sigma'} \langle\langle c^+_{\mathbf{q}\sigma'} c_{\mathbf{k}+\mathbf{q}\sigma'}; n^+_{\mathbf{k}}\rangle\rangle\end{aligned} \tag{5.92}$$

from which we obtain the RPA result,

$$\langle\langle n_{\mathbf{k}}; n^+_{\mathbf{k}}\rangle\rangle = \frac{1}{2\pi} P_{\mathbf{k}}(\omega)\left\{1 - \left(\frac{4\pi e^2}{k^2\Omega}\right) P_{\mathbf{k}}(\omega)\right\}^{-1} \tag{5.93}$$

with $P_{\mathbf{k}}(\omega)$ given by (5.82). The corresponding expression for the dielectric function is,

$$\epsilon(\mathbf{k}, \omega) = 1 - (4\pi e^2/k^2\Omega)P_{\mathbf{k}}(\omega). \tag{5.94}$$

The screening of an applied frequency and wavevector dependent probe that couples to the charge density is proportional to the inverse of the dielectric function, i.e. an applied potential $V(\mathbf{k}, \omega)$ behaves as an effective potential $V(\mathbf{k}, \omega)/\epsilon(\mathbf{k}, \omega)$. For example, in calculating the lattice vibrational frequency spectrum of a metal the interatomic potential is screened by the conduction electrons, and since the Fermi energy is much larger than typical phonon energies the dielectric function in the effective, screened potential can be replaced by its static value to a good approximation. Singularities in the static dielectric function that occur when the wavevector spans parallel faces of the Fermi surface ($k = 2p_f$ for an isotropic, degenerate electron gas) modulate the phonon spectrum, and the resulting anomalies in the phonon dispersion curves were predicted by Kohn (1959).

If the RPA result for the dielectric function is substituted in Eq. (5.70) for the interaction energy, and the identity (5.71) used to calculate the total

ISBN 0-8053-6610-5, 0-8053-6611-3 (pbk.)

energy per electron of a degenerage gas the result is the sum of (5.89) and the terms

$$0.0622 \ln r_s - 0.096 \quad \text{(Rhydbergs)}. \tag{5.95}$$

These terms are the leading ones in the expansion in r_s of the correlation energy per electron that is due to the charge-induced correlations in the electron motion. Couched slightly differently, the correlation energy is the difference between the exact and Hartree-Fock energies.

Let us now study the equation

$$\operatorname{Re} \epsilon(\mathbf{k}, \omega) = 0, \tag{5.96}$$

for the collective modes of the electron gas. Because (5.94) is valid for small k we expand $P_{\mathbf{k}}(\omega)$ in k. For $k < p_f$ and $\omega > p_f k/m^*$ we find,

$$P_{\mathbf{k}}(\omega) \simeq \frac{k^2}{m^* \omega^2} \sum_{\mathbf{p}\sigma} f_p = \frac{k^2 N}{m^* \omega^2} \cdot \tag{5.97}$$

Hence, to the same approximation

$$\epsilon(\mathbf{k}, \omega) = 1 - \omega_p^2/\omega^2 \tag{5.98}$$

where the plasmon frequency*

$$\omega_p = (4\pi e^2 n/m^*)^{1/2}. \tag{5.99}$$

We conclude that for small k the electron gas supports a collective oscillation of the particle density with a frequency ω_p. Notice that the collective mode frequency is finite for $k \to 0$ because of the long-range nature of the Coulomb potential. If the expansion (5.97) is taken to the next order in k^2 the dispersion of the collective mode is found to be

$$\{\omega_p^2 + \tfrac{3}{5}E_f(2k^2/m^*)\}^{1/2}. \tag{5.100}$$

We have written the result in a way which emphasizes that the coefficient of k^2 is proportional to the average kinetic energy of an electron.

The dispersion of the mode that satisfies (5.96) is sketched in Fig. (5.2). Notice that for a given wavevector there are two values of ω which satisfy Eq. (5.96). The lower-frequency solution is not physically significant, however, since it occurs within the particle-hole continuum and, in consequence, it is heavily damped and gives no significant structure to the dissipative part of the Green's function. The plasmon contribution to the Green's function has been observed in several types of experiment, including electron scattering, and the interested reader is referred to the review

---

* For simple metals ω_p is of the order of several electron volts, e.g. for sodium, $\omega_p = 5.7$ eV, whereas for a semiconductor $\omega_p = (4\pi n e^2/m^* \epsilon_0)^{1/2}$ is often only a few milli-electron volts.

ISBN 0-8053-6610-5, 0-8053-6611-3 (pbk.)

by Platzman and Wolff (1973) for further details. These authors also discuss the effect of band structure in the interpretation of experimental data.

Recent measurements of $S(k, \omega)$ by inelastic X-ray scattering demonstrate that, for the metals Li, Al and Be, the simple RPA described here is quantitatively inadequate in the interpretation of the data [Eisenberger et al. (1975), and references therein]. The qualitative features observed in $S(k, \omega)$ are in accord with the RPA; namely, for small k a plasmon excitation is observed, while at large k particle-hole excitations entirely dominate the spectrum, cf. Fig. (5.2). However, for intermediate values of k there are essentially two components to the spectrum, as illustrated in Fig. (5.3) for Al with $k = 1.59 p_f$. The low-frequency, plasmon contribution persists, in disagreement with the predictions of the RPA, and the high-frequency component, centered around $k^2/2m^*$, dominates the spectrum at larger k. The discrepancy between RPA and the data is believed to be due to the finite lifetime of the excitations, and this can be included by making

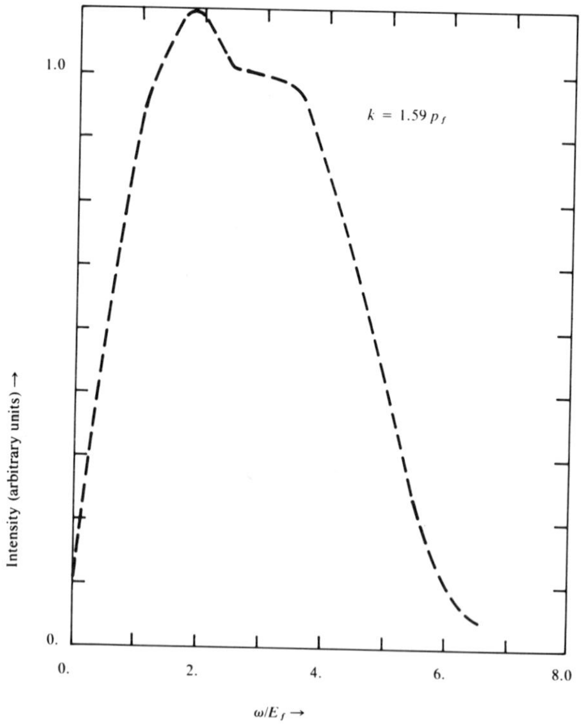

Fig. (5.3) The intensity of inelastically scattered X-rays is shown for Al as a function of ω (in units of E_f) for an intermediate wavevector $k = 1.59\ p_f$. [After Platzman and Eisenberger (1974).]

ISBN 0-8053-6610-5, 0-8053-6611-3 (pbk.)

a calculation along he lines of the following treatment of nonlinear spin waves. Another viable line of attack is to use a generalized Langevin equation, described in Chap. 3, which has proved very successful in the description of the dynamic properties of classical monatomic liquids.

5.4 SPIN WAVE INTERACTIONS IN A FERROMAGNET

The magnetic properties of a large number of insulating compounds are well described by the Heisenberg exchange Hamiltonian [Mattis (1965) and Keffer (1967)]

$$\mathcal{H} = - \sum_{\mathbf{l}\mathbf{l}'} J(\mathbf{l} - \mathbf{l}')\mathbf{S}_{\mathbf{l}} \cdot \mathbf{S}_{\mathbf{l}'} . \tag{5.101}$$

In (5.101), $J(\mathbf{l} - \mathbf{l}')$ is the exchange parameter for spins $\mathbf{S}_{\mathbf{l}}$ and $\mathbf{S}_{\mathbf{l}'}$ situated at the lattice's sites $\mathbf{l}$ and $\mathbf{l}'$, and it is defined such that $J(0) = 0$. A detailed comparison between experiment and theory based on (5.101) shows that small anisotropy terms must often be added to the Hamiltonian to obtain fully satisfactory agreement.* However, we shall not consider such terms in the following discussion.

Below a critical temperature, determined mainly by the magnitude of the exchange parameter [Stanley (1971)], we assume that the spins align ferromagnetically, i.e. the average value of $\langle S_{\mathbf{l}}^z \rangle$, say, is the same for all spins. Such a state is energetically favored if there is just one exchange parameter in (5.101) which is positive. At absolute zero the spins are fully aligned along the quantization axis, and $\langle S^z \rangle = S$, the magnitude of the spin.

Spin waves are collective modes of the system that create deviations from the fully aligned state, and they can be thermally excited, leading eventually to the complete loss of magnetization at the critical temperature, or created by an external probe such as in neutron scattering experiments [Squires (1978)].

Deviations from the fully aligned state are generated by the spin raising and lowering operators $S^{\pm} = S^x \pm iS^y$ which satisfy the commutation relations

$$[S_{\mathbf{l}}^+, S_{\mathbf{l}'}^-] = \delta_{\mathbf{l}\mathbf{l}'} 2S_{\mathbf{l}}^z \tag{5.102}$$

and

$$[S_{\mathbf{l}}^z, S_{\mathbf{l}'}^{\pm}] = \pm \delta_{\mathbf{l}\mathbf{l}'} S_{\mathbf{l}}^{\pm} . \tag{5.103}$$

---

* If dipolar terms are added to (5.101), for example, then the total spin in the direction of quantization is no longer a constant-of-motion, as we shall assume it to be in our discussion. These terms are important for the interpretation of nuclear magnetic resonance data and negligible in most neutron scattering experiments.

ISBN 0-8053-6610-5, 0-8053-6611-3 (pbk.)

In terms of these operators the Hamiltonian (5.101) is

$$\mathcal{H} = -\sum_{\mathbf{l}\mathbf{l}'} J(\mathbf{l}-\mathbf{l}')(S^z_{\mathbf{l}} S^z_{\mathbf{l}'} + S^+_{\mathbf{l}} S^-_{\mathbf{l}'})$$

and the equation-of-motion for S^+, say, is

$$i\dot{S}^+_{\mathbf{l}} = 2\sum_{\mathbf{l}'} J(\mathbf{l}-\mathbf{l}')(S^z_{\mathbf{l}'} S^+_{\mathbf{l}} - S^+_{\mathbf{l}'} S^z_{\mathbf{l}}). \tag{5.104}$$

In obtaining (5.104) we have used the fact that

$$J(\mathbf{l}-\mathbf{l}') = J(\mathbf{l}'-\mathbf{l}) \tag{5.105}$$

which is valid for simple crystal structures, including, of course, Bravais lattices. If the temperature of the system is small compared to the critical temperature, we can linearize the equation-of-motion (5.104) by replacing S^z by its maximum value, S. The resulting equation can be solved by introducing operators $S^\pm_{\mathbf{k}}$ through the canonical transformation

$$S^\pm_{\mathbf{l}} = \frac{1}{N}\sum_{\mathbf{k}} \exp(\pm i\mathbf{k}\cdot\mathbf{l}) S^\pm_{\mathbf{k}} \tag{5.106}$$

where N is the total number of spins, which is equal to the number of unit cells in the crystal in the present case. Introducing,

$$J(\mathbf{k}) = \sum_{\mathbf{l}} J(\mathbf{l}) \exp(-i\mathbf{k}\cdot\mathbf{l}) = J(-\mathbf{k}) \tag{5.107}$$

the equation-of-motion for $S^+_{\mathbf{k}}$ can be written

$$i\dot{S}^+_{\mathbf{k}} = 2S\{J(0) - J(\mathbf{k})\}S^+_{\mathbf{k}} \tag{5.108}$$

so that the Heisenberg operator

$$S^+_{\mathbf{k}}(t) = \exp(-i\omega_{\mathbf{k}} t) S^+_{\mathbf{k}} \tag{5.109}$$

where the spin wave dispersion,*

$$\omega_{\mathbf{k}} = 2S\{J(0) - J(\mathbf{k})\}. \tag{5.110}$$

The spin wave dispersion can be measured over a wide range of wavevectors by inelastic magnetic neutron scattering. Values of the exchange constants can then be found by fitting theoretical expressions for the dispersion to the measured dispersion. Figure (5.4) shows the spin wave dispersion in ferromagnetic terbium measured by inelastic neutron scattering. Because this rare earth metal has two atoms per unit cell there are "acous-

* If a magnetic field is included in (5.101) through a term $-h\sum S^z_{\mathbf{l}}$ a term h is added to the spin wave dispersion (5.110), i.e. a steady magnetic field introduces a gap in the spectrum at zero wavevector.

ISBN 0-8053-6610-5, 0-8053-6611-3 (pbk.)

tic'' and ''optic'' spin wave branches. The data can be interpreted in terms of a spin wave dispersion derived from a Hamiltonian which is the sum of a Heisenberg exchange term and complicated anisotropy terms. One effect of these anisotropy terms is to produce gaps in the dispersion at $\mathbf{k} = 0$. Even though terbium is a metal, the use of a localized electron Hamiltonian is justified since the magnetic moment on the terbium atoms, due to unpaired 4f electrons, is compactly distributed; in fact, the average radial distribution of the moment distribution is approximately 0.4Å which is much smaller than the interatomic spacing. Interactions between the spin waves and phonons lead to hybridization of the two collective modes as shown in Fig. (5.4).

From (5.110) it is at once obvious that he spin wave frequency goes to zero in the long-wavelength limit. For small values of k,

$$J(\mathbf{k}) = J(0) + k^2(D/2S) + \cdots$$

where D is the spin wave stiffness constant, and

$$\omega_{\mathbf{k}} \simeq Dk^2. \tag{5.111}$$

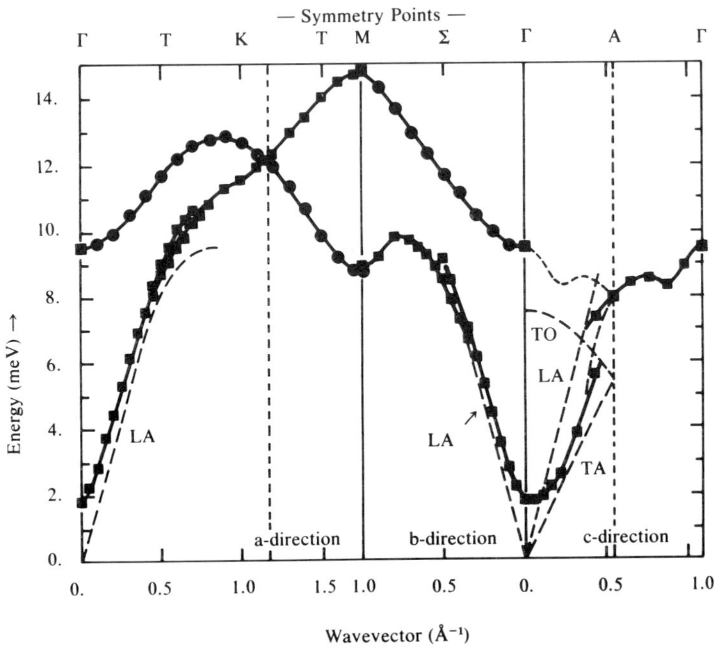

Fig. (5.4) The spin wave dispersion in terbium, measured by inelastic neutron scattering at 4.2°K, is shown for wavevectors along high-symmetry directions in the hexagonal close packed structure. The hybridization of spin waves and phonons is also included (LA = longitudinal acoustic, TA = transverse acoustic and TO = transverse optic phonon.) [After Mackintosh and Møller (1972).]

ISBN 0-8053-6610-5, 0-8053-6611-3 (pbk.)

For a nearest-neighbor exchange coupling of strength J,

$$D = 2SJa^2 \tag{5.112}$$

where a is the lattice parameter if the crystal is simple, body-centered, or face-centered cubic. EuO and EuS are simple ferromagnets in which the Eu^{2+} ions form a face-centered lattice. For EuO, $J = 0.052$ and the next nearest coupling is 0.013 meV.

When using the linear equation-of-motion (5.108) it is consistent to replace S^z in the commutation relation (5.102) by S. Hence, in this approximation, the operators $S_{\mathbf{k}}^{\mp}$ obey Bose statistics, and

$$\langle S_{\mathbf{k}}^{-} S_{\mathbf{k}}^{+} \rangle = 2SNn_{\mathbf{k}} \tag{5.113}$$

where the Bose distribution function

$$n_{\mathbf{k}} = \{\exp(\beta\omega_{\mathbf{k}}) - 1\}^{-1}. \tag{5.114}$$

The decrease in the magnetization from the saturated value S per spin due to thermal excitation of spin waves is given by (5.113). To obtain the precise form for the magnetization, i.e. $\langle S^z \rangle$, we need an expression for S^z in terms of $S^{\pm}$. Because we are interested in the low-temperature limit, where at most only one spin deviation is thermally excited, an expression valid for $S^z = S$ and $S^z = S - 1$ is adequate. It can be shown that he appropriate expression is

$$S^z = S - \frac{1}{2S} S^{-} S^{+}. \tag{5.115}$$

This approximate result and the linearized form of (5.102) preserve the commutation relations (5.103), and (5.115) is exact for the special case of $S = \frac{1}{2}$. From (5.115),

$$\begin{aligned} \langle S^z \rangle &= S - \frac{1}{2SN} \sum_{\mathbf{k}} \langle S_{\mathbf{k}}^{-} S_{\mathbf{k}}^{+} \rangle \\ &= S - \frac{v_0}{(2\pi)^3} \int d\mathbf{k}\, n_{\mathbf{k}} \end{aligned} \tag{5.116}$$

where v_0 is the volume of a unit cell. For low temperatures we can use the approximate result (5.111) for $\omega_{\mathbf{k}}$, and extend the range of integration in (5.116) over all $\mathbf{k}$ space with negligible error. Hence,

$$S - \langle S^z \rangle \simeq \frac{v_0}{4\pi^2} \left(\frac{k_B T}{D} \right)^{3/2} \int_0^\infty \frac{dx\, x^{1/2}}{(e^x - 1)} \tag{5.117}$$

and the integral is, apart from a constant, a Riemann zeta function. The result (5.117) tells us that the magnetization decreases with temperature as

ISBN 0-8053-6610-5, 0-8053-6611-3 (pbk.)

$T^{3/2}$. If we retain the next term in (5.111), proportional to k^4, then this introduces an additional term in (5.117) proportional to $T^{5/2}$.

The first detailed analysis of spin wave interactions was made by Dyson (1956). He separated the interaction of spin waves into two components. The kinematic interaction is a consequence of spin statistics, namely that the maximum number of spin deviations that can occur on any one spin is $2S$. The interaction that prevents this limit being exceeded is, clearly, a repulsive one. The second component of the interaction is attractive, and it arises because if the neighboring spins to which a given spin is exchange-coupled deviate from the fully aligned state, it costs less energy to create a deviation on the given spin than if all its neighbors are fully aligned along the axis of quantization. Dyson showed that the effect of the kinematic interaction is small compared to the dynamic interaction in the limit of low temperatures. Moreover, he showed that the effect of the dynamic interaction is described by the effective Hamiltonian*,

$$\mathscr{H} = \sum_{\mathbf{k}} \omega \mathbf{k} a_{\mathbf{k}}^{+} a_{\mathbf{k}} - \frac{1}{N} \sum_{1234} a_1^{+} a_2^{+} a_3 a_4 \delta_{1+2,3+4} \{J(1-3) - J(3)\} \qquad (5.118)$$

where $\omega_{\mathbf{k}}$ is given by (5.110), and the Bose operators a, a^+ satisfy

$$[a_{\mathbf{p}}, a_{\mathbf{q}}^{+}] = \delta_{\mathbf{pq}} \qquad (5.119)$$

with

$$S_{\mathbf{k}}^{z} = NS\delta_{\mathbf{k},0} - \sum_{\mathbf{p}} a_{\mathbf{k}+\mathbf{p}}^{+} a_{\mathbf{p}}. \qquad (5.120)$$

In (5.118) we have used a shorthand notation for wavevector, $\mathbf{k}_1 = 1$, etc., for brevity.

For nearest-neighbor exchange coupling, to which we shall restrict our attention,

$$J(\mathbf{k}) = rJ\gamma_{\mathbf{k}} \qquad (5.121)$$

where r is the number of neighbors, J is the magnitude of the coupling and $\gamma_{\mathbf{k}}$ is a geometric factor that depends on the crystal structure. For a body-centered cubic lattice, for example, of side a,

$$\gamma_{\mathbf{k}} = \cos(\tfrac{1}{2}ak_x)\cos(\tfrac{1}{2}ak_y)\cos(\tfrac{1}{2}ak_z),$$

and $r = 8$.

The first term in (5.118) describes linear spin waves, and the second, quartic term describes their interactions. We study the effect of the inter-

* The calculation of the collisional self-energy for the Dyson effective Hamiltonian and the alternative Hamiltonian derived from the Holstein-Primakoff transformation are compared, in the low temperature limit, by Rastelli and Lindgård (1979).

ISBN 0-8053-6610-5, 0-8053-6611-3 (pbk.)

actions by calculating the causal Green's function*,

$$G_{pq}(\omega) = \langle\langle a_p; a_q^+ \rangle\rangle \tag{5.122}$$

from its equation-of-motion

$$-\omega G_{pq}(\omega) = \frac{1}{2\pi}\langle [a_p, a_q^+]\rangle + \langle\langle [a_p, \mathscr{H}]; a_q^+ \rangle\rangle \tag{5.123}$$

where $\mathscr{H}$ is given by (5.118). If we neglect the quartic term in $\mathscr{H}$, for the moment, we obtain the noninteracting Green's function,

$$G_{pq}(\omega) = -\frac{1}{2\pi}\delta_{pq}(\omega + \omega_q)^{-1}. \tag{5.124}$$

This Green's function possesses a pole at $\omega = -\omega_p$ where ω_p is the spin wave dispersion, $\omega_p = 2rJS(1 - \gamma_p)$.

The exact result for the Green's function on the right-hand side of (5.123) is†

$$\langle\langle [a_p, \mathscr{H}]; a_q^+ \rangle\rangle = \omega_p G_{pq}(\omega) - \frac{rJ}{N}\sum_{234} \delta_{p+2,3+4} \times \{\gamma_{p-3} - 2\gamma_3 + \gamma_{2-3}\}\langle\langle a_2^+ a_3 a_4; a_q^+ \rangle\rangle. \tag{5.125}$$

Our first approximation for the interaction term on the right-hand side of this result is to linearize it by making a pairing approximation for the product

$$a_2^+ a_3 a_4 \simeq \overbrace{a_2^+ a_3} \underbrace{a_2^+ \, a_4} = \delta_{23}\langle a_2^+ a_2 \rangle a_4 + \delta_{24}\langle a_2^+ a_2 \rangle a_3 \tag{5.126}$$

and

$$\langle a_k^+ a_k \rangle = \{\exp(\beta\omega_k) - 1\}^{-1} = n_k. \tag{5.127}$$

From (5.125) and (5.126),

$$-\{\omega + \omega_k(T)\}G_{kp} = \frac{1}{2\pi}\delta_{kp} \tag{5.128}$$

where the temperature dependent spin wave frequency,

$$\omega_k(T) = 2rJS(1 - \gamma_k)\{1 - e(T)/S\} \tag{5.129}$$

---

* The two spin wave spectrum is described by the Green's function $\langle\langle a_1 a_2; a_3^+ a_4^+ \rangle\rangle$. Evaluated in the pair approximation described here, it gives the exact two spin wave bound-state dispersion, (in the limit of low temperatures) cf. Mattis (1965).

† In the following we shall often denote wavevector subscripts as scalar quantities, particularly when using the shorthand $\mathbf{k}_1 = 1$, etc.

ISBN 0-8053-6610-5, 0-8053-6611-3 (pbk.)

and $e(T)$ is the average spin wave energy

$$e(T) = \frac{1}{N} \sum_{\mathbf{q}} (1 - \gamma_{\mathbf{q}}) n_{\mathbf{q}}. \tag{5.130}$$

In obtaining (5.129) we have utilized the identity

$$\sum_{\mathbf{p}} \gamma_{\mathbf{k}-\mathbf{p}} n_{\mathbf{p}} = \gamma_{\mathbf{k}} \sum_{\mathbf{p}} \gamma_{\mathbf{p}} n_{\mathbf{p}}. \tag{5.131}$$

The first-order effect of the dynamic spin wave interaction is therefore to renormalize the spin wave energy by a factor $\{1 - e(T)/S\}$, which decreases with increasing temperature, i.e. the spin wave frequencies soften with increasing temperature. (5.129) and (5.130) form a pair of equations for $\omega_{\mathbf{k}}(T)$ that must be solved self-consistently, since $n_{\mathbf{q}}$ in (5.130) is the Bose distribution function for the frequency. In practice it is found that a self-consistent solution exists, for three-dimensional models, up to some temperature that is quite close to the critical temperature estimated by sophisticated series expansion methods. A detailed comparison by Dietrich *et al.* (1976) beween theory and measurements of the spin wave dispersion in EuO shows that the simple renormalization scheme is adequate over a

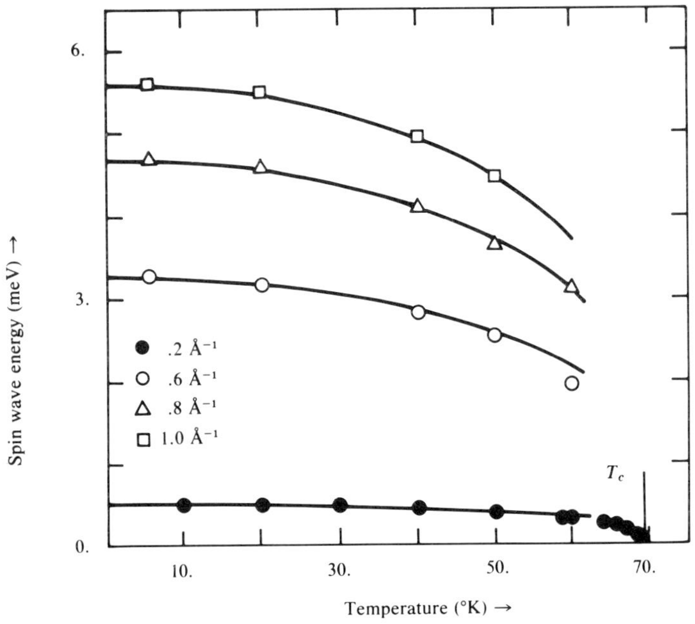

Fig. (5.5) Measured spin wave energies in EuO for various wavevectors, normalized to their values at 5°K, are shown as a function of temperature. The critical temperature $T_c = 69.1$°K. [After Dietrich et al. (1976).]

ISBN 0-8053-6610-5, 0-8053-6611-3 (pbk.)

wide range of temperatures and some of their results are shown in Fig. (5.5). Here we focus on the low-temperature regime where the theory is expected to be most reliable. For small k, we obtain from (5.129),

$$\omega_{\mathbf{k}}(T) \simeq k^2 D(T) \tag{5.132}$$

which defines a temperature dependent spin wave stiffness constant $D(T)$. Using (5.132) in (5.130), and expanding $(1 - \gamma_{\mathbf{q}})$ to the first nonvanishing order, leads to

$$\begin{aligned} e(T) &\simeq \left(\frac{a^2}{rN}\right) \sum_{\mathbf{q}} q^2\{\exp(\beta D(0) q^2) - 1\}^{-1} \\ &= \left(\frac{a^2 v_0}{2\pi^2 r}\right) \int_0^\infty dq\, q^4\{\exp(\beta D(0) q^2) - 1\}^{-1}, \end{aligned} \tag{5.133}$$

from which we conclude that $e(T) \propto T^{5/2}$. We have therefore found that the dynamic interaction between the spin waves gives, to leading order, a temperature dependence to D such that it decreases with increasing temperature, T, as $T^{5/2}$. The magnetieation can be calculated from (5.120). The dynamic interaction is found to give a contribution to the magnetization that leads to a decrease in the magnetization with temperature as T^4, in accord with the findings of Dyson (1956).

To go beyond the result (5.128), and obtain an expression for the spin wave damping, we write the Green's function on the right-hand side of (5.125)

$$\begin{aligned} \langle\langle a_2^+ a_3 a_4; a_q^+ \rangle\rangle &= \langle\langle \overline{a_2^+ a_3} a_4; a_q^+ \rangle\rangle + A(2, 3, 4) \\ &= n_2\{\delta_{23} G_{4q} + \delta_{24} G_{3q}\} + A(2, 3, 4) \end{aligned} \tag{5.134}$$

and calculate $A(2, 3, 4)$ to first order (we do not display the dependence of A on $\mathbf{q}$ for notational convenience). Substituting (5.134) in (5.125), and using the result in (5.123) gives

$$\begin{aligned} -\{\omega + \omega_k(T)\} G_{kp}(\omega) = \frac{1}{2\pi} \delta_{kp} - \left(\frac{rJ}{N}\right) \sum_{234} \delta_{k+2,3+4} \\ \times \{\gamma_{k-3} - 2\gamma_3 + \gamma_{2-3}\} A(2, 3, 4). \end{aligned} \tag{5.135}$$

An equation for $A(2, 3, 4)$ is formed by constructing the equation-of-motion for $\langle\langle a_2^+ a_3 a_4; a_q^+ \rangle\rangle$, which involves a lot of tedious algebra. Defining

$$E(p, q, q') = \omega_q + \omega_{q'} - \omega_p \tag{5.136}$$

ISBN 0-8053-6610-5, 0-8053-6611-3 (pbk.)

the result is,

$$\{\omega + E(p, q, q')\}[n_p\{\delta_{pq}G_{q'k} + \delta_{pq'}G_{qk}\} + A(p, q, q')]$$
$$= -\frac{1}{2\pi}\langle[a_p^+ a_q a_{q'}, a_k^+]\rangle + \frac{rJ}{N}\sum_{1234}\delta_{1+2,3+4}$$
$$\times [\{\gamma_{1-3} - 2\gamma_3 + \gamma_{2-3}\}\{\delta_{1q}\langle\langle a_p^+ a_2^+ a_{q'} a_3 a_4; a_k^+\rangle\rangle$$
$$+ \delta_{1q'}\langle\langle a_p^+ a_q a_2^+ a_3 a_4; a_k^+\rangle\rangle\}$$
$$- \delta_{3p}\{\gamma_{1-3} - \gamma_3 + \gamma_{1-4} - \gamma_4\}\langle\langle a_1^+ a_2^+ a_4 a_q a_{q'}; a_k^+\rangle\rangle]. \quad (5.137)$$

The Green's functions G_{qk} and $G_{q'k}$ that appear here on the left-hand side are transferred to the right-hand side of the equation and expressed in terms of fourth-order Green's functions $\langle\langle a_2^+ a_3 a_4; a_k^+\rangle\rangle$ through the equation-of-motion (5.123) and (5.125). Having done this we break up the first two sixth-order Green's functions in (5.137) and observe that the fourth-order Green's functions coming from G_{qk} and $G_{q'k}$ are exactly cancelled by components from the decoupled sixth-order Green's functions. As a result of this cancellation the nonlinear term in the equation for $A(p, q, q')$ is

$$\frac{rJ}{N}\sum_{123}\delta_{1+2,p+3}\{\gamma_{1-p} + \gamma_{1-3} + \gamma_{2-p} + \gamma_{2-3} - 2\gamma_p - 2\gamma_3\}$$
$$\times \{n_p[\delta_{1q'}\langle\langle a_q a_2^+ a_3; a_k^+\rangle\rangle + \delta_{1q}\langle\langle a_2^+ a_{q'} a_3; a_k^+\rangle\rangle]$$
$$- \tfrac{1}{2}\langle\langle a_1^+ a_2^+ a_3 a_q a_{q'}; a_k^+\rangle\rangle\}. \quad (5.138)$$

The fourth- and sixth-order Green's functions on the right-hand side of (5.138) are approximated in terms of G's, the lowest-order Green's functions. This gives a first-order approximation for $A(p, q, q')$, as required. There is an extensive cancellation of terms, and, in particular, the inhomogeneous term in (5.137) is exactly cancelled by the corresponding contributions from G_{qk} and $G_{q'k}$. The final result is

$$\{\omega + E(p, q, q')\}A(p, q, q')$$
$$= \left(\frac{2rJ}{N}\right) G_{q+q'-k,k}\{n_p(1 + n_q + n_{q'}) - n_q n_{q'}\}$$
$$\times \{\gamma_{p-q} + \gamma_{p-q'} - \gamma_p - \gamma_{q+q'-p}\}. \quad (5.139)$$

It now remains to substitute (5.139) in (5.135) for the Green's function $G_{\mathrm{kp}}(\omega)$. The result can be written

$$G_{\mathrm{kp}}(\omega) = -\frac{1}{2\pi}\delta_{\mathrm{kp}}\{\omega + \omega_{\mathrm{k}}(T) + \sum_{\mathrm{k}}(\omega)\}^{-1} \quad (5.140)$$

ISBN 0-8053-6610-5, 0-8053-6611-3 (pbk.)

where the collisional self-energy,

$$\sum_{\mathrm{k}}(\omega) = -\left(\frac{2r^2J^2}{N}\right)\sum_{123}\delta_{k+1,2+3}\{n_1(1+n_2+n_3)-n_2n_3\}$$
$$\times\{\gamma_{q-2}+\gamma_{q-3}-\gamma_2-\gamma_3\}\{\gamma_{q-2}+\gamma_{q-3}-\gamma_1-\gamma_q\}$$
$$\times\{\omega+E(1,2,3)\}^{-1}. \quad (5.141)$$

A result of particular interest is that the self-energy has a nonvanishing complex component when we make the substitution $\omega \rightarrow \omega - i\eta$ and $\eta \rightarrow 0$. This component represents a lifetime for the spin wave of wavevector k, and the denominator of the Green's function is,

$$G_{\mathrm{kk}}^{-1}(\omega) \propto \left\{-\omega - \omega_{\mathrm{k}}(T) - \Delta(\mathbf{k},\omega) + \frac{i}{2}\Gamma(\mathbf{k},\omega)\right\} \quad (5.142)$$

where

$$\Gamma(\mathbf{k},\omega) = -2\lim_{\eta\rightarrow 0}\operatorname{Im}\sum_{\mathrm{k}}(\omega - i\eta). \quad (5.143)$$

and $\Delta(\mathbf{k},\omega)$ is the corresponding real part of the self-energy.

A first-order estimate of the lifetime is obtained from (5.143), with the result (5.141) for the self-energy, by setting $\omega = -\omega_{\mathrm{k}}$. We evaluate the resulting expression in the limit of low temperatures and small $\mathbf{k}$. If we assume that $\omega_{\mathrm{k}} \gg k_BT$ the products of Bose occupation factors in (5.141) can be neglected, to a good approximation. We then find,

$$\Gamma_{\mathrm{k}} = \Gamma(\mathbf{k},\omega_{\mathrm{k}}) = \left(\frac{2\pi rJ}{N^2}\right)\sum_{\mathrm{pq}} n_{\mathrm{p}}\delta(\gamma_{\mathrm{q}}+\gamma_{\mathrm{k+p-q}}-\gamma_{\mathrm{k}}-\gamma_{\mathrm{p}})$$
$$\times\{\gamma_{\mathrm{k-q}}+\gamma_{\mathrm{p-q}}-\gamma_{\mathrm{k}}-\gamma_{\mathrm{p}}\}^2. \quad (5.144)$$

At low temperatures the Bose factor n_{p} makes the dominant contribution from the sum over $\mathbf{p}$ come from small $\mathbf{p}$. Because we have specified that $\mathbf{k}$ is small the delta function restricts $\mathbf{q}$ to be small also. Hence we can expand the γ functions in terms of their arguments. For example,

$$\gamma_{\mathrm{p-q}} \simeq 1 - \tfrac{1}{6}a^2(p^2+q^2-2\mathbf{p}\cdot\mathbf{q})$$

for a simple cubic lattice of side a. With this type of approximation,

$$\Gamma_{\mathrm{k}} \simeq \left(\frac{4\pi Ja^2}{SN^2}\right)\sum_{\mathrm{pq}} n_{\mathrm{p}}(\mathbf{k}\cdot\mathbf{p})^2\delta\{q^2-\mathbf{q}\cdot(\mathbf{k}+\mathbf{p})+\mathbf{k}\cdot\mathbf{p}\}$$
$$= \left(\frac{Ja^2v_0}{2\pi SN}\right)\sum_{\mathrm{p}} n_{\mathrm{p}}(\mathbf{k}\cdot\mathbf{p})^2|\mathbf{k}-\mathbf{p}|, \quad (5.145)$$

where in arriving at the last line we have used the result, valid for three-

ISBN 0-8053-6610-5, 0-8053-6611-3 (pbk.)

dimensional crystals,

$$\frac{1}{N}\sum_{\mathbf{q}} \delta\{q^2 - \mathbf{q}\cdot(\mathbf{k} + \mathbf{p}) + \mathbf{k}\cdot\mathbf{p}\} = \frac{v_0}{8\pi^2}\,|\mathbf{k} - \mathbf{p}|. \qquad (5.146)$$

From (5.145) it is at once evident that the spin wave damping is proportional to k^3 at long wavelengths and low temperatures (recall that we assume $\omega_{\mathbf{k}} \gg k_B T$). Moreover, it is a simple matter to prove that the temperature dependence is given by $T^{5/2}$; hence, the important result is $\Gamma_{\mathbf{k}} \propto k^3 T^{5/2}$ in the limit of low temperatures and long wavelengths, whereas in the hydrodynamic regime, discussed in Sec. 3.6, $\Gamma_{\mathbf{k}} \propto k^4$.

Most magnetic salts form an antiferromagnetic structure below the critical temperature in which neighboring ions align in opposite directions, e.g. MnF_2 and $RbMnF_3$. The temperature renormalization of the spin wave dispersion can be carried through with minimal added complexity compared to the calculation given here for a ferromagnet. However, the calculation of the spin wave damping for an antiferromagnet is very complicated. A comparative study of theory and experimental results for $RbMnF_3$ is given by Rezende and White (1978).

ISBN 0-8053-6610-5, 0-8053-6611-3 (pbk.)

CHAPTER VI

LINEAR RESPONSE THEORY

The usefulness of linear response theory to study the dynamic properties of matter rests on the relation established in Sec. 1.5 between the change in a variable induced by a small-time dependent perturbation and the dynamic susceptibility. In consequence, we can obtain the dynamic susceptibility, and thereby the spectrum of spontaneous fluctuations, by calculating the change induced in the average value of the relevant operator caused by a time-dependent perturbation. For the sake of completeness we record here the pertinent results obtained in the discussion given in Sec. 1.5.

The starting point is to modify the Hamiltonian which describes the system of interest by a small perturbation

$$- A^{+}h(t) \tag{6.1}$$

where the scalar strength function $h(t)$ satisfies $h(-\infty) = 0$, and A^{+} is the Hermitian conjugate of the relevant variable. If $h(t)$ is chosen to have the specific time dependent ($\eta \to 0$)

$$h(t) = h_0 \exp(i\omega t + \eta t) \tag{6.2}$$

where h_0 and ω are constants, then the change in the average value of the variable A, to first order in h_0, is *

$$h(t)\tilde{\chi}_A(i\omega + \eta), \tag{6.3}$$

where $\tilde{\chi}_A$ is the dynamic susceptibility for the variable A. We could equally well use a discontinuous perturbation in place of (6.1), in which case $\tilde{\chi}_A$ in (6.3) would be replaced by the relaxation function introduced in Sec. 1.6.

If the probe, described by the Hamiltonian (6.1), couples to a collective oscillation of the unperturbed system than $\tilde{\chi}_A\ (i\omega)$ will display a resonance

---

* We attach a suffix A to $\tilde{\chi}$ for convenience in subsequent sections.

Stephen Lovesey, Condensed Matter Physics: Dynamic Correlations

ISBN 0-8053-6610-5, 0-8053-6611-3 (pbk.)

when ω coincides with the collective mode frequency. The spectrum of spontaneous fluctuations in A can also be obtained from a knowledge of $\tilde{\chi}_A(i\omega)$ by using the fluctuation-dissipation theorem, Eq. (1.50); this theorem states that the imaginary, or dissipative, part of the dynamic susceptibility

$$\tilde{\chi}''_A(i\omega) = -\tfrac{1}{2}\{1 - \exp(-\omega\beta)\} \int_{-\infty}^{\infty} dt \exp(-i\omega t)\langle A^+A(t)\rangle \tag{6.4}$$

where the inverse temperature $\beta = 1/k_B T$.

Hence, if we calculate, using perturbation theory for example, the change in the variable A induced by the probe defined by the Hamiltonian (6.1) then the dynamic susceptibility is obtained, according to (6.3), by dividing the induced change by the scalar strength function. We shall illustrate this procedure in the subsequent sections by studying the properties of an electron gas subject to a steady magnetic field, and the collective spin excitations in the Hubbard model of an itinerant ferromagnet. The reader who is particularly interested in the properties of disordered materials is urged to study the calculation by Hubbard and Beeby (1969) of the phonon spectrum of a disordered material.

6.1 ELECTRON GAS IN A STEADY MAGNETIC FIELD

The dynamic properties of a homogeneous electron gas in zero field are discussed in Sec. 5.3. It is well known that the motion of an electron is changed substantially by a magnetic field; the field dependent electron states are usually referred to as Landau orbitals, or states, following his pioneering work on the diamagnetic susceptibility of an electron gas [Landau (1930); for a review see, for example, Abrikosov (1972)]. Here we choose to concentrate on the effect of a magnetic field on the wavevector and frequency dependent dielectric function and spin susceptibility.

We begin by reviewing the nature of the Landau states for an electron of mass m^* and charge $-e$. Let a steady magnetic field of strength H act along the z direction. The energy of an electron is then specified by its wavevector component p_z and a positive integer n, and it is [Landau and Liftshitz (1965), Abrikosov (1972)]

$$E_\nu = (p_z^2/2m^*) + (n + \tfrac{1}{2})\omega_c \tag{6.5}$$

where the cyclotron frequency* (c = velocity of light)

$$\omega_c = (eH/m^*c). \tag{6.6}$$

In (6.5), ν denotes the quantum numbers p_z and n, and it is evident that the energy is the sum of the kinetic energy of free motion along the z axis and

* A useful formula for ω_c, with H in units of 10^4 gauss, is $\omega_c = (0.116\, H/(m^*/m_e))$meV, where m_e is the electron mass.

ISBN 0-8053-6610-5, 0-8053-6611-3 (pbk.)

an isotropic harmonic oscillator with a characteristic frequency ω_c. Hence, in the presence of a magnetic field two of the momentum quantum numbers of a free electron are replaced in the energy eigenvalue by a positive integer which labels harmonic oscillator states. The counting of states is a little tricky, and we shall return to this problem after describing the spatial wave function associated with the energy (6.5).

It is convenient to introduce a wavevector $\theta = (m^* \omega_c)^{1/2}$ which is the inverse of the radius of the lowest energy Landau orbital. Note that θ is independent of the effective mass m^*. The electrons are assumed to form a parallelepiped with sides of length L_1, L_2 and L_3, and so $\Omega = L_1L_2L_3$. The spatial dependence of the νth normalized wavefunction depends on the quantum numbers p_y, p_z and n and it is*

$$|\nu\rangle = (\theta/L_2L_3)^{1/2} \exp(iyp_y + izp_z)h_n(\theta x - p_y/\theta), \tag{6.7}$$

where h_n is a normalized Hermite function of order n. The functions $h_n(x)$ satisfy the orthogonality condition,

$$\int_{-\infty}^{\infty} dx\, h_n(x)h_m(x) = \delta_{nm}. \tag{6.8}$$

The allowed values of p_y and p_z are obtained by imposing periodic boundary conditions on $|\nu\rangle$ with period L_2 and L_3. The range of p_y is confined to the interval $0 \le p_y \le \theta^2 L_1$ by the restriction imposed by the boundary condition along the x-axis [Peierls (1955) and Abrikosov (1972)]. The integer n can take any nonnegative intergral value, as we have already remarked. These results on the allowed values of the quantum numbers p_y, p_z and n in (6.5) and (6.7) mean that a sum over the quantum number ν, which is shorthand for all three quantum numbers, becomes

$$\sum_\nu = \frac{L_2L_3}{(2\pi)^2} \sum_{n=0}^{\infty} \int_0^{\theta^2 L_1} dp_y \int_{-\infty}^{\infty} dp_z. \tag{6.9}$$

The Fermi distribution function for the νth state depends explicitly on the spin state of the electron. We label spin states by the index σ which takes two values, ± 1, indicated sometimes by $\uparrow$ and $\downarrow$ to denote components parallel and antiparallel to the applied magnetic field. The Fermi distribution function is then ($\nu \equiv p_y, p_z, n$ and σ)

$$f_\nu = (\exp\{\beta(E_{p_z n} - \mu_\sigma)\} + 1)^{-1} \tag{6.10}$$

and

$$\mu_\sigma = \mu + \sigma \tfrac{1}{4}\, g\, \omega_c(m^*/m_e) \tag{6.11}$$

* The vector potential has components $A_y = Hx$, and $A_x = A_z = 0$. Later we shall choose our external wavevector **k** such that the x-axis is perpendicular to the plane of **H** and **k**.

ISBN 0-8053-6610-5, 0-8053-6611-3 (pbk.)

where μ is the chemical potential and g is the electron gyromagnetic ratio. We have chosen to display the values of the gyromagnetic ratio and mass, m^*, so that the results we obtain can be used to describe properties of free carriers in doped semiconductors where g and m^* can differ significantly from the free electron values $g = 2$ and $m^* = m_e$. For example, with the narrow band semiconductor n-InSb it is possible to have $g \sim -50$ and $m^* \sim 0.015\ m_e$. Platzman and Wolff (1973) review a variety of phenomena that have been observed in semiconductors in a steady magnetic field.

For moderate values of the field the magnetic energy imparted to an electron, ω_c, is usually very small compared to the Fermi energy. In the case of nearly free, degenerate electrons of density* n_0 in a simple metal, the Fermi energy

$$E_f = (p_f^2/2m^*), \quad \text{with } p_f = (3\pi^2 n_0)^{1/3} \tag{6.12}$$

is of the order of several eV, and this is to be compared with the value $\omega_c = 1.16$ meV for $H = 100$ kG which is a moderately large magnetic field. When the condition $E_f \gg \omega_c$ is satisfied for a degenerate electron gas the chemical potential μ in (6.11) is given by,

$$\mu = E_f \tag{6.13}$$

to a very good approximation [Greene *et al.* (1969)].

The condition $E_f \gg \omega_c$ can be reversed in the case of doped semiconductors since n_0 can be several orders of magnitude smaller than for electrons in simple metals and often $m^* \ll m_e$. When n_0 and m^* are small, a moderate magnetic field acts as an intense perturbation, $E_f \ll \omega_c$, and μ is no longer determined solely by the density of charge carriers. Under these conditions the electrons occupy the lowest energy Landau level and their spins align parallel or antiparallel to the applied field depending on the sign of the gyromagnetic ratio. It is instructive to look at the calculation of the chemical potential μ in some detail since the mathematical steps involved are similar to those we shall need in more complicated calculations later on in the section.

The chemical potential is obtained from the thermodynamic potential F using the relation [Landau and Lifshitz (1959)]

$$n_o = -\frac{1}{\Omega}\left(\frac{\partial F}{\partial \mu}\right) \tag{6.14}$$

* We use n_0 rather than n to denote the electron density to avoid possible confusion with the quantum number n.

ISBN 0-8053-6610-5, 0-8053-6611-3 (pbk.)

and the result

$$F = -\frac{1}{\beta}\sum_{\nu} \ln[1 + \exp\{\beta(\mu_\sigma - E_{p_z n})\}]$$

$$= -\frac{\theta^2\Omega}{\beta 4\pi^2}\sum_{\sigma}\sum_{n}\int_{-\infty}^{\infty} dp_z \ln[1 + \exp\{(b_{n\sigma}^2 - p_z^2)\beta/2m^*\}] \quad (6.15)$$

where the wavevector

$$b_{n\sigma} = \{2m^*(\mu - (n + \tfrac{1}{2})\omega_c + \gamma\omega_c\sigma)\}^{1/2} \quad (6.16)$$

and

$$\gamma = \tfrac{1}{4}\, g(m^*/m_e). \quad (6.17)$$

In obtaining the second equality in (6.15) we have used (6.9). The result (6.15) can be integrated by parts, and, writing $p_z = p$, we arrive at the more useful form

$$F = -\frac{\theta^2\Omega}{2\pi^2 m^*}\sum_{\sigma}\sum_{n}\int_{0}^{\infty} dp\ p^2[1+ \exp\{(p^2 - b_{n\sigma}^2)\beta/2m^*\}]^{-1}. \quad (6.18)$$

In the limit of low temperatures, $\beta \to \infty$, the range of integration in (6.18) is restricted so that the argument of the exponential is negative and for a given n and σ this requires $p \leq b_{n\sigma}$. The low-temperature limit of (6.18) is therefore [Abrikosov (1972), Chap. 10]

$$F = -\frac{\theta^2\Omega}{2\pi^2 m^*}\sum_{\sigma}\sum_{n} \tfrac{1}{3}\, b_{n\sigma}^3 \quad (6.19)$$

Taking the limit of a large field, $\omega_c \to \infty$, it is evident that the significant terms in (6.19) are those with $n = 0$, and $\sigma = 1$ or -1 depending on the sign of γ. Hence the required result for the thermodynamic potential is

$$F = -\frac{\theta^2\Omega}{6\pi^2 m^*}\{2m^*(\mu - \tfrac{1}{2}\,\omega_c + |\gamma|\omega_c)\}^{3/2}. \quad (6.20)$$

Note that for free electrons $|\gamma| = \frac{1}{2}$. Using (6.20) in (6.14) we obtain a result for μ valid for low temperatures and large applied fields.

$$\mu = \omega_c\{b^2/2m^*\omega_c + \tfrac{1}{2} - |\gamma|\} \quad (6.21)$$

where the wavevector

$$b = 2\pi^2 n_o/m^*\omega_c, \quad (6.22)$$

now plays a role similar to that of the Fermi wavevector p_f in the case of a moderate applied field. Substituting (6.21) into (6.16),

$$b_{n\sigma} = \{2m^*(b^2/2m^* - n\omega_c + (\sigma\gamma - |\gamma|)\omega_c)\}^{1/2} \quad (6.23)$$

ISBN 0-8053-6610-5, 0-8053-6611-3 (pbk.)

from which it is clear that the minimum value of $b_{n\sigma}$ as a function of n and σ is b, and this is achieved for $n = 0$ and the appropriate choice of σ.

Having discussed briefly the properties of Landau states we turn to the study of the dynamic properties of an electron gas in a magnetic field. If the variable of interest A is a single particle operator we can use the second quantized representation [Schiff (1955) and Landau and Lifshitz (1965)]

$$A = \sum_{\lambda\lambda'} A_{\lambda\lambda'} c^+_\lambda c_{\lambda'} \tag{6.24}$$

where c^+_λ, c_λ are the Fermi operators for the Landau state $|\lambda\rangle$, and $A_{\lambda\lambda'}$ is the matrix element of A taken between the states $|\lambda\rangle$ and $|\lambda'\rangle$,

$$A_{\lambda\lambda'} = \langle\lambda|A|\lambda'\rangle. \tag{6.25}$$

The Hamiltonian for the states $|\lambda\rangle$ is

$$\mathcal{H} = \sum_{\lambda} E_\lambda c^+_\lambda c_\lambda \tag{6.26}$$

with E_λ given by the sum of (6.5) and the spin dependent energy, $-\sigma\gamma\omega_c$.

From (6.24) it is clear that we can find the change in the average value of A due to the perturbation $-h(t)A^+$ if we know the change in the operator* $c^+_\mu c_\nu$. Using the equation-of-motion

$$i\partial_t c^+_\mu c_\nu = [c^+_\mu c_\nu, \mathcal{H} - A^+ h(t)], \tag{6.27}$$

we find [the commutators can be evaluated with the aid of (5.79)]

$$i\partial_t c^+_\mu c_\nu = (E_\nu - E_\mu) c^+_\mu c_\nu + h(t) \sum_\lambda (A^*_{\mu\lambda} c^+_\lambda c_\nu - A^*_{\lambda\nu} c^+_\mu c_\lambda). \tag{6.28}$$

Integrating this equation, and displaying the time arguments of the operators and using the convention $c^+_\mu(0) = c^+_\mu$,

$$\begin{aligned} c^+_\mu(t) c_\nu(t) &= \exp\{it(E_\mu - E_\nu)\} c^+_\mu c_\nu \\ &\quad + i \exp\{it(E_\mu - E_\nu)\} \int_{-\infty}^{t} d\bar{t} \exp\{i\bar{t}(E_\nu - E_\mu)\} h(\bar{t}) \\ &\quad \sum_\lambda (A^*_{\lambda\nu} c^+_\mu(\bar{t}) c_\lambda(\bar{t}) - A^*_{\mu\lambda} c^+_\lambda(\bar{t}) c_\nu(\bar{t})). \end{aligned} \tag{6.29}$$

Because we work to first order in $h(t)$, the time dependence of the operators involved in the second term in (6.29) is taken to be the same as that in the unperturbed system. i.e. the first term in (6.29). With $h(t)$ given by (6.2) we

* Note that μ is used here as a quantum number, and its use for this purpose should hopefully not cause confusion with the chemical potential.

ISBN 0-8053-6610-5, 0-8053-6611-3 (pbk.)

obtain from (6.29) the result,

$$
\begin{aligned}
c_\mu^+(t)c_\nu(t) = \exp\{it(E_\mu - E_\nu)\}c_\mu^+c_\nu \\
&+ ih(t)\sum_\lambda \exp\{it(E_\mu - E_\nu)\} \\
&\times\left[\frac{A_{\lambda\nu}^* c_\mu^+ c_\lambda \exp\{it(E_\nu - E_\lambda)\}}{(i\omega + \eta + E_\nu - E_\lambda)}\right. \\
&\left. - \frac{A_{\mu\lambda}^* c_\lambda^+ c_\nu \exp\{it(E_\lambda - E_\mu)\}}{(i\omega + \eta + E_\lambda - E_\mu)}\right].
\end{aligned}
\tag{6.30}
$$

We now take the average value of both sides of (6.30) with respect to the unperturbed states, and since, for example,

$$\langle c_\mu^+ c_\lambda \rangle = \delta_{\mu\lambda} f_\lambda \tag{6.31}$$

where f_λ is the Fermi distribution function (6.10), we obtain the change in the average value of $c_\mu^+ c_\nu$ to first order in $h(t)$,

$$-\,h(t)A_{\mu\nu}^*\Gamma_{\mu\nu}(\omega) \tag{6.32}$$

where the quantity $\Gamma_{\mu\nu}$ is

$$\Gamma_{\mu\nu}(\omega) = (f_\nu - f_\mu)/(\omega - i\eta + E_\nu - E_\mu). \tag{6.33}$$

It follows from this result, and the definition of the dynamic susceptibility (6.3), that the noninteracting susceptibility is

$$\tilde{\chi}_A(i\omega) = -\sum_{\mu\nu} |A_{\mu\nu}|^2 \Gamma_{\mu\nu}(\omega). \tag{6.34}$$

Before proceeding to discuss the effect of the Coulomb interaction on the susceptibility we use the result (6.34) to obtain the noninteracting electron density susceptibility in zero field, and verify that the result is identical with the one obtained in the previous chapter using causal Green's functions. Hence, we take

$$
\begin{aligned}
A = n_{\mathbf{k}} &= \sum_j \exp(-i\mathbf{k}\cdot\mathbf{R}_j) \\
&= \sum_{\mathbf{p}\sigma}\sum_{\mathbf{q}\sigma'} \langle \mathbf{p}\sigma|\exp(-i\mathbf{k}\cdot\mathbf{R})|\mathbf{q}\sigma'\rangle c_{\mathbf{p}\sigma}^+ c_{\mathbf{q}\sigma'}.
\end{aligned}
\tag{6.35}
$$

The plane-wave states

$$|\mathbf{p}\rangle = \Omega^{-1/2}\exp(i\mathbf{p}\cdot\mathbf{R}) \tag{6.36}$$

and the matrix element

$$
\begin{aligned}
A_{\mu\nu} &\equiv \langle \mathbf{p}\sigma|\exp(-i\mathbf{k}\cdot\mathbf{R})|\mathbf{q}\sigma'\rangle \\
&= ((2\pi)^3/\Omega)\,\delta_{\sigma\sigma'}\delta(\mathbf{p} + \mathbf{k} - \mathbf{q}).
\end{aligned}
\tag{6.37}
$$

ISBN 0-8053-6610-5, 0-8053-6611-3 (pbk.)

The density of wavevectors is $\Omega/(2\pi)^3$, and so we obtain from (6.35) together with (6.37)

$$n_{\mathbf{k}} = \sum_{\mathbf{p}\sigma} c^+_{\mathbf{p}\sigma} c_{\mathbf{k}+\mathbf{p}\sigma} \tag{6.38}$$

in agreement with (5.76). The noninteracting, electron density susceptibility is obtained from (6.33) and (6.34) using (6.37), with the result

$$\tilde{\chi}_n(\mathbf{k},i\omega) = \sum_{\mathbf{p}\sigma} (f_p - f_{k+p})/(\omega - i\eta + E_{k+p} - E_p), \tag{6.39}$$

which agrees with the result for the Green's function, $-(1/2\pi)\tilde{\chi}_n(\mathbf{k},i\omega)$, Eq. (5.82).

We now address the problem of accounting for the Coulomb interaction in the calculation of the susceptibility for the variable A. The main effect of the probe, $-h(t)A^+$, is, of course, to induce a fluctuation in the average value of A. In addition, however, the probe can create a fluctuation in the electron density. Hence, the total perturbation to which the electron gas responds is the sum of the applied perturbation and the potential due to the induced electron density fluctuation. We denote the induced density fluctuation by δn, and the electrostatic potential V associated with the fluctuation is determined by Poisson's equation,

$$\nabla^2 V = -4\pi e \delta n. \tag{6.40}$$

The induced potential will have the same space- and time-dependence as the applied perturbation. Hence, if the applied perturbation varies harmonically in space with a wavevector $\mathbf{k}$ [as it does in the example (6.35)] and the time-dependence is as given in (6.2), then the Fourier transform of the electrostatic potential is, from (6.40),

$$V_{\mathbf{k}} = h_0 \delta n_{\mathbf{k}} (4\pi e/k^2) \exp(i\omega t + \eta t), \tag{6.41}$$

and the total perturbation to which the electron gas is subjected is

$$-h(t)A^+_{\mathbf{k}} + h(t)\delta n_{\mathbf{k}}(4\pi e^2/k^2\Omega) n^+_{\mathbf{k}}, \tag{6.42}$$

where, for completeness, we have added a subscript $\mathbf{k}$ to the variable that describes the applied perturbation. The change in the average value of $c^+_\mu c_\nu$ brought about by the perturbation (6.41) is, from (6.32),

$$h(t)\Gamma_{\mu\nu}(\omega)\{-(A_{\mathbf{k}})^*_{\mu\nu} + \delta n_{\mathbf{k}}\left(\frac{4\pi e^2}{k^2\Omega}\right)\langle\mu|\exp(-i\mathbf{k}\cdot\mathbf{R})|\nu\rangle^*\}. \tag{6.43}$$

Hence, we find the dynamic susceptibility for processes described by the variable $A_{\mathbf{k}}$ by multiplying (6.43) by $\Sigma(A_{\mathbf{k}})_{\mu\nu}$ and dividing the result by $h(t)$,

ISBN 0-8053-6610-5, 0-8053-6611-3 (pbk.)

and the answer is,

$$\tilde{\chi}_A(\mathbf{k},i\omega) = -\sum_{\mu\nu} |(A_{\mathbf{k}})_{\mu\nu}|^2\Gamma_{\mu\nu}(\omega)$$

$$+ \delta n_{\mathbf{k}}\left(\frac{4\pi e^2}{k^2\Omega}\right)\sum_{\mu\nu}(A_{\mathbf{k}})_{\mu\nu}\langle\mu|\exp(-i\mathbf{k}\cdot\mathbf{R})|\nu\rangle^*\Gamma_{\mu\nu}(\omega). \qquad (6.44)$$

The first term in (6.44) is identical with the noninteracting susceptibility (6.34), apart from the fact that in (6.44) we display explicitly the wavevector dependence of the matrix element, $(A_{\mathbf{k}})_{\mu\nu}$.

It remains only to determine $\delta n_{\mathbf{k}}$ in the second term in (6.44). From (6.35) it follows that we can obtain $\delta n_{\mathbf{k}}$ by multiplying (6.43), which is the change in the average value of $c^+_\mu c_\nu$ due to the applied perturbation, by the matrix element

$$(N_{\mathbf{k}})_{\mu\nu} = \langle\mu|\exp(-i\mathbf{k}\cdot\mathbf{R})|\nu\rangle, \qquad (6.45)$$

and summing the result over the quantum numbers μ,ν; the result of this calculation gives us $h(t)\delta n_{\mathbf{k}}$. If we define the function

$$P_{\mathbf{k}}(\omega) = \sum_{\mu\nu}|(N_{\mathbf{k}})_{\mu\nu}|^2\Gamma_{\mu\nu}(\omega) \qquad (6.46)$$

we obtain the result

$$\delta n_{\mathbf{k}} = -\{1 - \left(\frac{4\pi e^2}{k^2\Omega}\right)P_{\mathbf{k}}(\omega)\}^{-1}\sum_{\mu\nu}(A_{\mathbf{k}})^*_{\mu\nu}(N_{\mathbf{k}})_{\mu\nu}\Gamma_{\mu\nu}(\omega). \qquad (6.47)$$

the susceptibility for the processes described by the variable $A_{\mathbf{k}}$, namely

$$\tilde{\chi}_A(\mathbf{k},i\omega) = -\sum_{\mu\nu}|(A_{\mathbf{k}})_{\mu\nu}|^2\Gamma_{\mu\nu}(\omega)$$

$$-\left(\frac{4\pi e^2}{k^2\Omega}\right)\left\{1 - \left(\frac{4\pi e^2}{k^2\Omega}\right)P_{\mathbf{k}}(\omega)\right\}^{-1}$$

$$\times \{\sum_{\mu\nu}(A_{\mathbf{k}})_{\mu\nu}(N_{\mathbf{k}})^*_{\mu\nu}\Gamma_{\mu\nu}(\omega)\}\{\sum_{\mu\nu}(A_{\mathbf{k}})^*_{\mu\nu}(N_{\mathbf{k}})_{\mu\nu}\Gamma_{\mu\nu}(\omega)\}. \qquad (6.48)$$

As a first example we apply (6.48) to the important problem of density fluctuations in an electron gas. In this instance $A_{\mathbf{k}} \equiv n_{\mathbf{k}}$, and the matrix elements $(A_{\mathbf{k}})_{\mu\nu} = (N_{\mathbf{k}})_{\mu\nu}$. With the definition (6.46), we obtain from (6.48)

ISBN 0-8053-6610-5, 0-8053-6611-3 (pbk.)

the result,

$$\tilde{\chi}_n(\mathbf{k},i\omega) = -P_{\mathbf{k}}(\omega) - \left(\frac{4\pi e^2}{k^2\Omega}\right)\left\{1 - \left(\frac{4\pi e^2}{k^2\Omega}\right)P_{\mathbf{k}}(\omega)\right\}^{-1} P_{\mathbf{k}}^2(\omega)$$

$$= -P_{\mathbf{k}}(\omega)\left\{1 - \left(\frac{4\pi e^2}{k^2\Omega}\right)P_{\mathbf{k}}(\omega)\right\}^{-1}. \tag{6.49}$$

When we recall that the corresponding causal Green's function is $-\tilde{\chi}_n(\mathbf{k},i\omega)/2\pi$, the result (6.49) has exactly the same form as the RPA Green's function for an electron gas in zero magnetic field, Eq. (5.93). The dielectric function is

$$\epsilon(\mathbf{k},\omega) = 1 - (4\pi e^2/k^2\Omega)P_{\mathbf{k}}(\omega). \tag{6.50}$$

We have already shown that the present formulation reproduces the unperturbed Green's function for an electron gas in zero field, for which the states $|\nu\rangle$ are the plane-wave states (6.36).

Hence, the result of the present section sheds a different light on the nature of the RPA approximation discussed in Chap. V, where it was the outcome of a certain decoupling scheme. Here we see that the RPA is equivalent to treating the electron gas as a noninteracting electron gas in its response to the total perturbation, which is the sum of the applied and induced electrostatic perturbation. The induced electrostatic perturbation is calculated self-consistently from the induced charge density through application of Poisson's equation. For this reason the RPA dielectric function (6.50) is often referred to as the self-consistent dielectric function.

We now consider the problem of fluctuations in the spin density

$$g \sum_j \delta(\mathbf{r} - \mathbf{R}_j)s_j^z \tag{6.51}$$

where s_j^z is the spin of the jth electron. It follows from (6.51) that the matrix element

$$(A_{\mathbf{k}})_{\mu\nu} = g\langle\mu|\exp(-i\mathbf{k}\cdot\mathbf{R})s^z|\nu\rangle$$

$$= \tfrac{1}{2}g\delta_{\sigma_\mu\sigma_\nu}\sigma_\mu(N_{\mathbf{k}})_{\mu\nu} \tag{6.52}$$

where it is understood that the matrix element $(N_{\mathbf{k}})_{\mu\nu}$ is calculated with the spatial wave functions (6.7). In what follows we separate the spin and spatial

ISBN 0-8053-6610-5, 0-8053-6611-3 (pbk.)

quantum numbers in μ,ν. For example,

$$\sum_{\mu\nu} \frac{|(A_{\mathrm{k}})_{\mu\nu}|^2(f_\nu - f_\mu)}{(\omega - i\eta + E_\nu - E_\mu)}$$

$$\Rightarrow \tfrac{1}{4}g^2 \sum_{\mu\sigma_\mu} \sum_{\nu\sigma_\nu} \frac{\delta_{\sigma_\mu\sigma_\nu}|(N_{\mathrm{k}})_{\mu\nu}|^2(f_{\nu\sigma_\mu} - f_{\mu\sigma_\mu})}{(\omega - i\eta + E_{\nu\sigma_\mu} - E_{\mu\sigma_\mu})}$$

$$= \tfrac{1}{4}g^2 \sum_{\mu\nu} \frac{|(N_{\mathrm{k}})_{\mu\nu}|^2}{(\omega - i\eta + E_\nu - E_\mu)} \{(f_{\nu\uparrow} - f_{\mu\uparrow}) + (f_{\nu\downarrow} - f_{\mu\downarrow})\}$$

$$= \tfrac{1}{4}g^2 \sum_{\mu\nu} |(N_{\mathrm{k}})_{\mu\nu}|^2 \{\Gamma^{\uparrow}_{\mu\nu}(\omega) + \Gamma^{\downarrow}_{\mu\nu}(\omega)\}$$

$$= \tfrac{1}{4}g^2\{P^{\uparrow}_{\mathrm{k}}(\omega) + P^{\downarrow}_{\mathrm{k}}(\omega)\}. \tag{6.53}$$

The penultimate and final equalities define, respectively, functions $\Gamma^{\sigma}_{\mu\nu}(\omega)$ and $P^{\sigma}_{\mathrm{k}}(\omega)$, and

$$P_{\mathrm{k}}(\omega) = P^{\uparrow}_{\mathrm{k}}(\omega) + P^{\downarrow}_{\mathrm{k}}(\omega) \tag{6.54}$$

where $P_{\mathrm{k}}(\omega)$ is the same function that appears in (6.46). Notice that the only quantities in these functions that depend on the spin index, σ, are the Fermi distribution functions, since the spin dependent energy terms in the denominator of $\Gamma^{\sigma}_{\mu\nu}(\omega)$ cancel. A similar calculation to one just given for the noninteracting spin susceptibility shows that

$$\sum_{\mu\nu} \frac{(A_{\mathrm{k}})_{\mu\nu}(N_{\mathrm{k}})^*_{\mu\nu}(f_\nu - f_\mu)}{(\omega - i\eta + E_\nu - E_\mu)} \Rightarrow \tfrac{1}{2}g \sum_{\mu\nu} \sum_{\sigma} \sigma|(N_{\mathrm{k}})_{\mu\nu}|^2\Gamma^{\sigma}_{\mu\nu}(\omega)$$

$$= \tfrac{1}{2}g\{P^{\uparrow}_{\mathrm{k}}(\omega) - P^{\downarrow}_{\mathrm{k}}(\omega)\}. \tag{6.55}$$

Using (6.53) and (6.54) in (6.48) shows that the spin susceptibility

$$\tilde{\chi}_s(\mathbf{k},i\omega) = -\tfrac{1}{4}g^2\{P^{\uparrow}_{\mathrm{k}}(\omega) + P^{\downarrow}_{\mathrm{k}}(\omega)\}$$
$$- \tfrac{1}{4}g^2(4\pi e^2/k^2\Omega\epsilon(\mathbf{k},\omega))$$
$$\times \{P^{\uparrow}_{\mathrm{k}}(\omega) - P^{\downarrow}_{\mathrm{k}}(\omega)\}^2. \tag{6.56}$$

In the absence of a field, $P^{\uparrow}_{\mathrm{k}}(\omega) = P^{\downarrow}_{\mathrm{k}}(\omega)$, and the second term in (6.56), due to the Coulomb interaction, vanishes. In other words, in zero field there is no coupling between the spin and particle densities.

The first step in evaluating the susceptibilities (6.49) and (6.56) is the calculation of the matrix element $(N_{\mathrm{k}})_{\mu\nu}$. In doing so we shall restrict attention to the case where **k** is perpendicular to the applied magnetic field, which defines the z-direction. Henceforth we shall not write the wavevector in vector notation, and without loss of generality we assume that it is parallel to the y-axis. From the definitions of $(N_{\mathrm{k}})_{\mu\nu}$ and the Landau states (6.7),

ISBN 0-8053-6610-5, 0-8053-6611-3 (pbk.)

and writing $\mu = \mathbf{p}, n$ and $\nu = \mathbf{q}, m$, we have

$$(N_\mathrm{k})_{\mu\nu} = \left(\frac{\theta}{L_2 L_3}\right) \int dx \int dy \int dz \exp(-iyp_y - izp_z) h_n(\theta x - p_y/\theta)$$
$$\times \exp(iyk) h_m(\theta x - q_y/\theta) \exp(iyq_y + izq_z). \tag{6.57}$$

If L_2 and L_3 are very large the intergrations over y and z result in delta functions with arguments $q_y + k - p_y$ and $q_z - p_z$, respectively. Remembering that the density of q_y and q_z vectors is $(2\pi)^2/L_2 L_3$ we obtain the following result for the square of the matrix element

$$|(N_\mathrm{k})_{\mu\nu}|^2 = \left(\frac{\theta}{L_2 L_3}\right)^2 (2\pi)^4 \delta(q_y + k - p_y)\delta(q_z - p_z) \frac{L_2 L_3}{(2\pi)^2}$$
$$\times \left\{\int_{-\infty}^{\infty} dx\, h_n(\theta x - p_y/\theta) h_m(\theta x - q_y/\theta)\right\}^2 \tag{6.58}$$

and so, from (6.9)

$$P_\mathrm{k}(\omega) = \left(\frac{L_2 L_3}{(2\pi)^2}\right)^2 \sum_\sigma \sum_n \int_0^{\theta^2 L_1} dp_y \int_{-\infty}^{\infty} dp_z \sum_m \int_0^{\theta^2 L_1} dq_y \int_{-\infty}^{\infty} dq_z \frac{(2\pi)^2 \theta^2}{L_2 L_3}$$
$$\times \delta(q_y + k - p_y)\delta(q_z - p_z)$$
$$\times \left\{\int_{-\infty}^{\infty} dx\, h_n(\theta x - p_y/\theta) h_m(\theta x - (p_y - k)/\theta)\right\}^2$$
$$\times (f_{mp_z\sigma} - f_{np_z\sigma})/(\omega - i\eta + (m - n)\omega_c)$$
$$= \left(\frac{\theta^2 \Omega}{4\pi^2}\right) \sum_\sigma \int_{-\infty}^{\infty} dp \sum_{mn} \left\{\theta \int_{-\infty}^{\infty} dx\, h_n(\theta x) h_m(\theta x + k/\theta)\right\}^2$$
$$\times (f_{mp\sigma} - f_{np\sigma})/(\omega - i\eta + (m - n)\omega_c). \tag{6.59}$$

Notice that because $q_z = p_z$ in the matrix element (6.58) the wavevector dependent components of the Landau energies cancel in the denominator. It now remains to evaluate the integral of the product of Hermite functions in (6.59).

Using the relation [Landau and Lifshitz (1965)]

$$h_m(\theta x + k/\theta) = \exp\left(\frac{k}{\theta^2}\frac{\partial}{\partial x}\right) h_m(\theta x)$$

the integral of interest can be written

$$Y_{nm}(k) = \theta \int_{-\infty}^{\infty} dx\, h_n(\theta x) \exp\left(\frac{k}{\theta^2}\frac{\partial}{\partial x}\right) h_m(\theta x). \tag{6.60}$$

ISBN 0-8053-6610-5, 0-8053-6611-3 (pbk.)

We now take advantage of the fact that $h_m(\theta x) = |m\rangle$ is the eigenfunction of a harmonic oscillator to write (6.60) in the appealing form,

$$Y_{nm}(k) = \langle n|\exp\left(\frac{ik}{\theta}p\right)|m\rangle, \tag{6.61}$$

where p is the momentum operator. In terms of Bose creation, and annihilation operators,

$$p = (-i/\sqrt{2})(a - a^+),$$

and the identity (1.26) lead to

$$\exp\left(\frac{ik}{\theta}p\right) = \exp\{(k/\theta\sqrt{2})(a - a^+)\}$$
$$= \exp(k^2/4\theta^2)\exp(ka/\theta\sqrt{2})\exp(-ka^+/\theta\sqrt{2}). \tag{6.62}$$

With the aid of the relation $|m\rangle = (m!)^{-1/2}(a^+)^m|0\rangle$, it is straightforward to show that

$$\exp\left(\frac{-ka^+}{\theta\sqrt{2}}\right)|m\rangle = \sum_{l=0}^{\infty}\left\{\frac{(l+m)!}{m!}\right\}^{1/2}\frac{1}{l!}\left(\frac{-k}{\theta\sqrt{2}}\right)^l|l+m\rangle. \tag{6.63}$$

Combining (6.62) and (6.63) in (6.61) we find that the latter is, for $n \geq m$,

$$Y_{nm}(k) = \exp(k^2/4\theta^2)\left(\frac{-k}{\theta\sqrt{2}}\right)^{n-m}(m!n!)^{-1/2}$$
$$\times\sum_{l=0}^{\infty}\left(\frac{-k^2}{2\theta^2}\right)^l\frac{(l+n)!}{l!\,(l+n-m)!} \tag{6.64}$$

and, in particular,

$$Y_{n0}(k) = (-k/\theta\sqrt{2})^n(n!)^{-1/2}\exp(-k^2/4\theta^2). \tag{6.65}$$

We shall now use (6.65) to obtain the intense field form of the RPA dielectric function.

When evaluating (6.59) in the limit of an intense field and low temperatures, recall that for a large applied field the electrons occupy the lowest Landau level and their spins align parallel or antiparallel to the field. As a result we can make the replacements

$$f_{mp\sigma} = \delta_{m,0}\delta_{\sigma,\uparrow}, \qquad f_{np\sigma} = \delta_{n,0}\delta_{\sigma,\uparrow}$$

and the integral over p is restricted to the range $-b$ to b, where b is given (6.22). Using the notation (6.61) for the matrix element in $P_{\mathbf{k}}(\omega)$, Eq. (6.59),

ISBN 0-8053-6610-5, 0-8053-6611-3 (pbk.)

and taking the intense field and low-temperature limit in the latter,

$$P_{\mathbf{k}}(\omega) = P_{\mathbf{k}}^{\uparrow}(\omega) = \left(\frac{\theta^2\Omega}{4\pi^2}\right)\int_{-b}^{b} dp \sum_{mn}^{\infty} \frac{|Y_{nm}(k)|^2(\delta_{m,0}-\delta_{n,0})}{(\omega - i\eta + (m-n)\omega_c)}$$

$$= 2N\omega_c \exp(-k^2/2\theta^2) \sum_{n=1}^{\infty} \frac{1}{(n-1)!}\left(\frac{k^2}{2\theta^2}\right)^n$$

$$\times \{(\omega - i\eta)^2 - (n\omega_c)^2\}^{-1}. \tag{6.66}$$

In obtaining the final result we have used (6.65).

The dissipative parts of the noninteracting density and spin susceptibilities are, according to (6.49) and (6.56), proportional to the imaginary part of $P_{\mathbf{k}}(\omega)$, Eq. (6.66). Some simple algebra shows that the imaginary part of $P_{\mathbf{k}}(\omega)$ is

$$\operatorname{Im} P_{\mathbf{k}}(\omega) = N\pi \exp(-k^2/2\theta^2) \sum_{n=1}^{\infty} \frac{1}{n!}\left(\frac{k^2}{2\theta^2}\right)^n$$

$$\times \{\delta(\omega - n\omega_c) - \delta(\omega + n\omega_c)\}. \tag{6.67}$$

Each term in the sum, $n = 1, 2, \cdots$, corresponds to the excitation, or de-excitation, of a state of energy $n\omega_c$, and the nth process has an amplitude

$$\exp(-k^2/2\theta^2)(1/n!)(k^2/2\theta^2)^n.$$

If $k \ll \theta$, only the first few terms in the sum are significant. Moreover, if we use (6.67) to calculate the density, or spin, autocorrelation function only one delta function will contribute to the result since $k_BT \ll \omega_c$.

In the absence of a magnetic field the electron density undergoes a collective oscillation with a frequency $\omega_p = (4\pi e^2 n_0/m^*)^{1/2}$ in the low-temperature and long-wavelength limit. The modification to this result brought about by an intense magnetic field can be found from the equation

$$1 = (4\pi e^2/k^2\Omega) \operatorname{Re} P_{\mathbf{k}}(\omega), \tag{6.68}$$

using (6.66) for $P_{\mathbf{k}}(\omega)$. Expanding (6.66) in powers of k/θ, and solving (6.68) to order $(k/\theta)^2$ the collective mode frequency is found to be ($\omega_p^2 \neq 3\omega_c^2$, $\mathbf{k} \perp \mathbf{H}$)

$$\omega_0^2 = \omega_p^2 + \omega_c^2 + \left(\frac{3k^2}{2m^*}\right)\left(\frac{\omega_c\omega_p^2}{\omega_p^2 - 3\omega_c^2}\right). \tag{6.69}$$

We see from this result that, in the long-wavelength limit, the collective mode frequency is shifted by the applied field from ω_p to $\{\omega_p^2 + \omega_c^2\}^{1/2}$. The functional form of the dispersion is the same as in the zero field case, and differs only by the fact that the coefficient of k^2 in (6.69) can be negative.

We now calculate the contribution that the collective mode makes to the dissipative part of the spin susceptibility. In the intense field limit the

ISBN 0-8053-6610-5, 0-8053-6611-3 (pbk.)

spin susceptibility (6.56) reduces to

$$\tilde{\chi}_s(k,i\omega) = -\tfrac{1}{4}g^2 P_k^{\uparrow}(\omega) - \tfrac{1}{4}g^2\left(\frac{4\pi e^2}{k^2\Omega\epsilon(k,\omega)}\right)\{P_k^{\uparrow}(\omega)\}^2. \tag{6.70}$$

Because the imaginary part of $P_k^{\uparrow}(\omega)$ is finite only when ω is a multiple of the cyclotron frequency, the first term in (6.70) does not contribute for $\omega = \omega_0$, where the collective mode frequency ω_0 is given by (6.69) to first order in k^2/θ^2. Therefore, for $\omega \sim \omega_0$,

$$\text{Im}\,\tilde{\chi}_s(k,i\omega) = -\left(\frac{4\pi e^2}{k^2\Omega}\right)\{\tfrac{1}{2}g\,\text{Re}\,P_k^{\uparrow}(\omega_0)\}^2\,\text{Im}\,\epsilon^{-1}(k,\omega). \tag{6.71}$$

The imaginary part of $1/\epsilon(k,\omega)$ is calculated by the method described at the end of Sec. 5.2 using the result, correct to lowest order in k/θ,

$$\epsilon(k,\omega) = 1 + \omega_p^2/(\omega_c^2 - \omega^2), \tag{6.72}$$

with the result,

$$\text{Im}\,\epsilon^{-1}(k,\omega) = \pi(\omega_p^2/2\omega_0)\{\delta(\omega-\omega_0) - \delta(\omega+\omega_0)\}. \tag{6.73}$$

Using this last result in conjunction with

$$\text{Re}\,P_k^{\uparrow}(\omega_0) \simeq N(k^2/m^*\omega_p^2), \tag{6.74}$$

we find that the collective mode contribution to the spin susceptibility is, in the intense-field and long-wavelength limit,

$$\text{Im}\,\tilde{\chi}_s(k,i\omega) = -N\pi\tfrac{1}{4}g^2(k^2/2m^*\omega_0)\{\delta(\omega-\omega_0) - \delta(\omega+\omega_0)\}. \tag{6.75}$$

The corresponding result for the density susceptibility is very easy to obtain from

$$\text{Im}\,\tilde{\chi}_n(k,i\omega) = -\text{Re}\,P_k(\omega_0)\,\text{Im}\,\epsilon^{-1}(k,\omega)$$

and the results (6.73) and (6.74). Assembling the terms we find,

$$\text{Im}\,\tilde{\chi}_n(k,i\omega) = -N\pi(k^2/2m^*\omega_0)\{\delta(\omega-\omega_0) - \delta(\omega+\omega_0)\}, \tag{6.76}$$

which is identical with the result (6.75) for the spin susceptibility apart from the factor $g^2/4$.

All the quantities of interest here, e.g. the RPA dielectric function, can be obtained from a knowledge of the function $P_k(\omega)$ which is, apart from a minus sign, the noninteracting electron density susceptibility. From (6.59) we have, for $\mathbf{k} \perp \mathbf{H}$,

$$\begin{aligned} P_k(\omega) &= P_k^{\uparrow}(\omega) + P_k^{\downarrow}(\omega) \\ &= N\left(\frac{\theta^2}{4\pi^2 n_0}\right)\sum_{\sigma}\int_{-\infty}^{\infty} dp \\ &\quad\times \sum_{mn} |Y_{nm}(k)|^2 (f_{mp\sigma} - f_{np\sigma})/(\omega - i\eta + (m-n)\omega_c), \end{aligned} \tag{6.77}$$

ISBN 0-8053-6610-5, 0-8053-6611-3 (pbk.)

and an explicit expression for the matrix element $Y_{nm}(k)$ is given in Eq. (6.64). The form for the matrix element that we have given is ideal for the study of the intense field limit where $E_f \ll \omega_c$ and the chemical potential is given by (6.21). However, it is not a convenient form for the moderate field limit, $E_f \gg \omega_c$ and $\mu = E_f$, since many terms contribute to the sum in (6.64). The algebra involved in obtaining a form for $Y_{nm}(k)$ for moderate fields and low temperatures is rather complicated and we shall not elaborate on it here. A detailed discussion is given by Greene *et al.* (1969), for example. We define a dimensionless parameter

$$\zeta_\sigma = k\,\{2\,(\mu - \sigma\gamma\omega_c)/m^*\}^{1/2}/\omega_c, \tag{6.78}$$

and following Greene *et al.* (1969) obtain the result, valid for $\mathbf{k} \perp \mathbf{H}$ and low temperatures,

$$P_k^\sigma(\omega) = -\frac{N3k^2}{2m^*\omega_c^2\zeta_\sigma^2}\left\{1 - g_0(\zeta_\sigma) + 2(\omega/\omega_c)^2 \times \sum_{n=1}^{\infty} \frac{g_n(\zeta_\sigma)}{n^2 + (\eta + i\omega/\omega_c)^2}\right\}, \tag{6.79}$$

where the function $g_n(\zeta\sigma)$ involves the integral of a Bessel function of order n, $J_n(x)$,

$$g_n(\zeta) = \int_0^{\pi/2} dx\, \sin x J_n^2(\zeta \sin x). \tag{6.80}$$

In deriving (6.79) we have neglected small terms that oscillate as a function of the applied field.

It is evident from (6.79) that the dissipative parts of the noninteracting density and spin susceptibilities are sums of delta functions $\delta(\omega - n\omega_c)$, just as we found for the intense field case, which is to be expected. The amplitude of the nth order process is $ng_n(\zeta_\sigma)$. For $n \geq 1$, the functions $g_n(\zeta)$ vanish for both small and large values of ζ. A simple calculation shows that for small ζ,

$$g_n(\zeta) \simeq \zeta^{2n}, \qquad n \geq 1 \tag{6.81}$$

and

$$g_0(\zeta) = 1 - \tfrac{1}{35}\zeta^2 + \cdots. \tag{6.82}$$

The reader can easily verify that the long-wavelength limit of the dielectric function for moderate fields is the same as for the intense field, (6.72). However there are differences in the higher-order terms in $\epsilon(k,\omega)$, and the collective mode frequency is now determined by ($\omega_p^2 \neq 3\omega_c^2$, $\mathbf{k} \perp \mathbf{H}$)

$$\omega_0^2 = \omega_p^2 + \omega_c^2 + \tfrac{3}{5}E_f\left(\frac{2k^2\omega_p^2/m^*}{\omega_p^2 - 3\,\omega_c^2}\right) \tag{6.83}$$

ISBN 0-8053-6610-5, 0-8053-6611-3 (pbk.)

and for zero field, $\omega_c = 0$, this agrees with the corresponding result derived in Sec. 5.3. The contribution this mode makes to the spin susceptibility (6.56) can be calculated following the line of argument used for the intense field limit. After calculating $P_k^{\uparrow} - P_k^{\downarrow}$ to order ζ^4, we find $(\omega_p^2 \neq 3\omega_c^2)$

$$\operatorname{Im} \tilde{\chi}_s(k, i\omega) = -N\pi\tfrac{1}{4} g^2 (k^2/m^*)^3 \frac{(3g\mu_B H/5)^2}{2\omega_0(\omega_p^2 - 3\omega_c^2)^2}$$

$$\times \{\delta(\omega - \omega_0) - \delta(\omega + \omega_0)\}, \tag{6.84}$$

where μ_B is the Bohr magneton. We have written (6.84) in a form in which the field dependence of the numerator is explicitly displayed, and we see that the amplitude vanishes as the field H goes to zero as H^2. The amplitude in (6.84) is very different from the corresponding amplitude in the intense field limit. The reason for this is that in the moderate field limit the amplitude of the collective mode contribution to the spin susceptibility is determined by the fluctuation in the spin density $P_k^{\uparrow} - P_k^{\downarrow}$, whereas in the intense field limit the spins are completely polarized either parallel or antiparallel to the field, depending on the sign of the gyromagnetic ratio.

6.2 COLLECTIVE SPIN MODES IN THE HUBBARD MODEL OF ITINERANT MAGNETISM

The development of a satisfactory theory of the magnetic properties of metals like iron and nickel is the subject of much research [Herring (1966) and Hubbard (1979)]. A basic problem is that some experimental data on these itinerant ferromagnets can be satisfactorily interpreted in terms of a localized electron model (Heisenberg exchange model discussed in Sec. 5.4) while other data are readily interpretable in terms of itinerant electron, or band theory. For example, how in the localized model can one understand the non-half-integral atomic moments observed? On the other hand, spin waves in ferromagnetic metals obey the fundamental predictions of the Heisenberg ferromagnet. The key to understanding these seemingly opposing characteristics of localized and itinerant electron states is believed to be the correlation between the electrons. For if, in spite of their band motion, the 3d electrons of an atom are strongly correlated with each other but only weakly with electrons on other atoms, Then such intra-atomic correlation leads inevitably to a behavior that is to some degree characteristic of an atomic, localized electron model.

Hubbard (1963) showed that electrons in narrow bands are described by a Hamiltonian in which the interaction is repulsive and operates between electrons of opposite spin on the same atom. We shall show that the cor-

ISBN 0-8053-6610-5, 0-8053-6611-3 (pbk.)

relation caused by this "on site" repulsion is that which leads to the possibility of collective spin modes, or spin waves.*

We shall first sketch a derivation of the Hubbard Hamiltonian for a single band of electrons, and then pass on to a discussion of the spin dynamics. The essential step in deriving the Hamiltonian is to express the Coulomb interaction between electrons in a narrow band in terms of operators $c^+_{\mathbf{l}\sigma}$ and $c_{\mathbf{l}\sigma}$ that create and destroy an electron in the state $\varphi(\mathbf{r} - \mathbf{l})$ centered about the lattice sites $\mathbf{l}$ [Peierls (1955) and Abrikosov (1972)]. The two spin states of an electron are labeled by $\sigma = \pm 1$, and sometimes it is convenient to write this in the form $\sigma = \uparrow$, or $\sigma = \downarrow$, to denote spin up and spin down electron states. The operators obey Fermi-Dirac statistics ($\mu \equiv \mathbf{l}\sigma$)

$$[c_\nu, c^+_\mu]_+ = \delta_{\mu\nu} \tag{6.85}$$

all other anticommutators being zero, and the particle number operator for the μth state, namely $n_\mu = c^+_\mu c_\mu$, obeys the relation

$$n_\mu n_\mu = n_\mu, \tag{6.86}$$

as a direct consequence of the condition

$$(c^+_\mu)^2 = (c_\mu)^2 = 0, \tag{6.87}$$

which is itself a direct consequence of the Pauli exclusion principle. A two-particle interaction A, like the Coulomb interaction between pairs of electrons, is of the general form

$$A = \sum_{i \neq j} a_{ij} \tag{6.88}$$

where the indices i and j label electrons, and the rule for second quantization is [Schiff (1955) and Landau and Lifshitz (1965)]

$$A = \tfrac{1}{2} \sum_{\mu\nu} \sum_{\mu'\nu'} \langle \varphi_\mu \varphi_\nu | a | \varphi_{\mu'} \varphi_{\nu'} \rangle c^+_\mu c^+_\nu c_{\nu'} c_{\mu'}. \tag{6.89}$$

Applied to the Coulomb interaction between electrons

$$e^2 \sum_{i \neq j} |\mathbf{R}_i - \mathbf{R}_j|^{-1} = \tfrac{1}{2} \sum_{\mathbf{lm}} \sum_{\mathbf{l}'\mathbf{m}'} \sum_{\sigma\sigma'} \langle \mathbf{lm} | R^{-1} | \mathbf{l}'\mathbf{m}' \rangle c^+_{\mathbf{l}\sigma} c^+_{\mathbf{m}\sigma'} c_{\mathbf{m}'\sigma'} c_{\mathbf{l}'\sigma} \tag{6.90}$$

where the matrix element

$$\langle \mathbf{lm} | R^{-1} | \mathbf{l}'\mathbf{m}' \rangle = e^2 \int d\mathbf{r} \int d\mathbf{r}' \, |\mathbf{r} - \mathbf{r}'|^{-1} \varphi^*(\mathbf{r} - \mathbf{l}) \varphi(\mathbf{r} - \mathbf{l}') \times \varphi^*(\mathbf{r}' - \mathbf{m}) \varphi(\mathbf{r}' - \mathbf{m}'). \tag{6.91}$$

* The discussion given here was suggested by Dr. J. Hubbard, and his development is given in unpublished lectures presented at an Advanced Nato Summer School on Magnetism, McGill University (1967), directed by Professor A. J. Freeman.

ISBN 0-8053-6610-5, 0-8053-6611-3 (pbk.)

In a narrow energy band the overlap of the functions φ centered about different sites is small. Therefore the largest of the matrix elemenents $\langle \mathbf{lm} | R^{-1} | \mathbf{l'm'} \rangle$ is that with $\mathbf{l} = \mathbf{m} = \mathbf{l'} = \mathbf{m'}$, and we shall denote this matrix element by U and neglect all others in (6.90). Hence the Coulomb interaction is approximated in the extreme narrow band or tight binding limit by

$$\tfrac{1}{2}U \sum_{\mathbf{l}} \sum_{\sigma\sigma'} c^+_{\mathbf{l}\sigma} c^+_{\mathbf{l}\sigma'} c_{\mathbf{l}\sigma'} c_{\mathbf{l}\sigma}. \tag{6.92}$$

Notice that all operators in (6.92) refer to the same atom.

The interaction (6.92) can be written in a more compact form by making the following manipulations of the Fermi operators. Starting from

$$c^+_\sigma c^+_{\sigma'} c_{\sigma'} c_\sigma = - c^+_\sigma c^+_{\sigma'} c_\sigma c_{\sigma'} = - c^+_\sigma (\delta_{\sigma\sigma'} - c_\sigma c^+_{\sigma'}) c_{\sigma'}$$

we have

$$\sum_{\alpha\lambda \gtrless} c^+_\sigma c^+_{\sigma'} c_{\sigma'} c_\sigma = - n_\uparrow - n_\downarrow + n_\uparrow n_\downarrow + n_\downarrow n_\downarrow + n_\uparrow n_\uparrow + n_\downarrow n_\uparrow$$

$$= n_\uparrow n_\downarrow + n_\downarrow n_\uparrow = \sum_\sigma n_\sigma n_{-\sigma}.$$

Using this last result in (6.92) we arrive at Hubbard's form for the interaction between electrons in an extremely narrow electron band, namely,

$$\tfrac{1}{2}U \sum_{\mathbf{l}\sigma} n_{\mathbf{l}\sigma} n_{\mathbf{l}-\sigma}. \tag{6.93}$$

The total Hamiltonian for electrons in a narrow band is the sum of (6.93) and the kinetic energy. If the creation and annihilation operators for the Bloch states $\varphi_{\mathbf{k}\sigma}$ are denoted by $c^+_{\mathbf{k}\sigma}$ and $c_{\mathbf{k}\sigma}$, with

$$c_{\mathbf{k}\sigma} = N^{-1/2} \sum_{\mathbf{l}} \exp(i\mathbf{k}\cdot\mathbf{l}) c_{\mathbf{l}\sigma}, \tag{6.94}$$

then the total Hamiltonian is

$$\mathscr{H} = \sum_{\mathbf{k}\sigma} E_k c^+_{\mathbf{k}\sigma} c_{\mathbf{k}\sigma} + \tfrac{1}{2}U \sum_{\mathbf{l}\sigma} n_{\mathbf{l}\sigma} n_{\mathbf{l}-\sigma}. \tag{6.95}$$

Here E_k is the band energy, and $E_k = E_{-k}$ by time-inversion symmetry.

As a first step toward understanding the properties of the system described by (6.95) we make a pairing approximation for the interaction term. That is to say, we take

$$n_{\mathbf{l}\sigma} n_{\mathbf{l}-\sigma} \simeq \langle n_{\mathbf{l}\sigma} \rangle n_{\mathbf{l}-\sigma} + n_{\mathbf{l}\sigma} \langle n_{\mathbf{l}-\sigma} \rangle - \langle n_{\mathbf{l}\sigma} \rangle \langle n_{\mathbf{l}-\sigma} \rangle. \tag{6.96}$$

The constant term in (6.96), can of course, be neglected in the present discussion. Inserting the first two terms in (6.96) into (6.95), and assuming that $\langle n_{\mathbf{l}-\sigma} \rangle$ is independent of the site position,

$$\mathscr{H} \simeq \sum_{\mathbf{k}\sigma} \{E_k + U\langle n_{-\sigma} \rangle\} c^+_{\mathbf{k}\sigma} c_{\mathbf{k}\sigma}. \tag{6.97}$$

ISBN 0-8053-6610-5, 0-8053-6611-3 (pbk.)

This Hamiltonian represents a band of noninteracting electrons with spin dependent energies $E_k + U\langle n_{-\sigma}\rangle$. It can be shown from (6.97) that ferromagnetic order exists if the Stoner condition for ferromagnetism is satisfied [Herring (1966)]. If the density of electron states at the Fermi energy is $Z(E_f)$ then the Stoner condition for ferromagnetic order is

$$UZ(E_f) > 1. \qquad (6.98)$$

When (6.98) is satisfied, the band splits into two bands, one of spin up electrons and one of spin down electrons separated by an energy $\Delta = U(\langle n_\downarrow\rangle - \langle n_\uparrow\rangle)$. In the paramagnetic regime, i.e. at a temperature above the critical temperature, $\langle n_\downarrow\rangle = \langle n_\uparrow\rangle$ and $\Delta = 0$.

We shall investigate the spin dynamics in the ferromagnetically-ordered state by calculating the susceptibility for the spin raising and lowering operators, which can also be regarded as the transverse spin components. The axis of quantization is taken to define the z axis. The spin angular momentum operators S^x and S^y when expressed in terms of raising and lowering operators S^+ and S^-, read

$$S^x = \frac{1}{2}(S^+ + S^-) \quad \text{and} \quad S^y = \frac{1}{2i}(S^+ - S^-).$$

The equivalent electron-spin raising and lowering operators (or spin-flip operators) are

$$S^+ = c_\uparrow^+ c_\downarrow \quad \text{and} \quad S^- = c_\downarrow^+ c_\uparrow, \qquad (6.99)$$

while the operator for its z component of spin is

$$S^z = \tfrac{1}{2}(c_\uparrow^+ c_\uparrow - c_\downarrow^+ c_\downarrow). \qquad (6.100)$$

These results are easily verified by calculating the spin **s** in second quantized form, i.e. by evaluating $\sum_{\sigma\sigma'} c_\sigma^+ \chi_\sigma^+ \mathbf{s} \chi_{\sigma'} c_{\sigma'}$ where χ_σ is the electron spinor. In view of (6.99) we shall study the dynamics of the operator $c_{\mathbf{k+q}\uparrow}^+ c_{\mathbf{q}\downarrow}$.

To this end we add to the Hamiltonian describing the electrons the interaction term ($\nu = +$ or $-$)

$$-\sum_{\mathbf{l}\nu} (S_{\mathbf{l}}^\nu)^+ h_{\mathbf{l}}^\nu(t) = -\sum_{\mathbf{l}\nu} S_{\mathbf{l}}^{-\nu} h_{\mathbf{l}}^\nu(t) \qquad (6.101)$$

where $h_{\mathbf{l}}^\nu(t)$ is the νth component of a magnetic field that varies in both space and time. For the moment we use the approximate Hamiltonian (6.97) to describe the electrons. Using the results (6.94) and (6.99), the total Hamiltonian given by the sum of (6.97) and (6.101) is

$$\mathcal{H} = \sum_{\mathbf{k}\sigma} \{E_k + U\langle n_{-\sigma}\rangle\} c_{\mathbf{k}\sigma}^+ c_{\mathbf{k}\sigma} - \frac{1}{N}\sum_{\mathbf{kq}} \{c_{\mathbf{k}\downarrow}^+ c_{\mathbf{k+q}\uparrow} h_{\mathbf{q}}^+ + c_{\mathbf{k+q}\uparrow}^+ c_{\mathbf{k}\downarrow} h_{\mathbf{q}}^-\}. \qquad (6.102)$$

ISBN 0-8053-6610-5, 0-8053-6611-3 (pbk.)

The equation-of-motion for $c^+_{k+q\uparrow}c_{q\downarrow}$ is obtained from (6.102) using (5.79), and the result is,

$$i\partial_t c^+_{k+q\uparrow}c_{q\downarrow} = [c^+_{k+q\uparrow}c_{q\downarrow}, \mathscr{H}]$$
$$= (E_q - E_{k+q} - \Delta)c^+_{k+q\uparrow}c_{q\downarrow}$$
$$-\frac{1}{N}\sum_p h^+_p(c^+_{k+q\uparrow}c_{q+p\uparrow} - c^+_{k+q-p\downarrow}c_{q\downarrow}). \quad (6.103)$$

With the choice

$$h^+_l(t) = h^+(t)\exp(i\,\mathbf{k}\cdot\mathbf{l})$$

we have

$$h^+_p(t) = N\delta_{kp}h^+(t). \quad (6.104)$$

Inserting this choice for $h^+_p(t)$ in the equation-of-motion (6.103) we have,

$$(i\partial_t + \Delta + E_{k+q} - E_q)c^+_{k+q\uparrow}(t)c_{q\downarrow}(t)$$
$$= -\,h^+(t)\{n_{k+q\uparrow}(t) - n_{q\downarrow}(t)\}, \quad (6.105)$$

where

$$n_{k\sigma} = c^+_{k\sigma}c_{k\sigma}. \quad (6.106)$$

We have displayed the time arguments in (6.105) for completeness.

To first order in the perturbing field h, $n_{k\sigma}$ is a constant independent of time. Because we require the solution of (6.105) correct to first order in h this means that in the second term the time dependent operators can be replaced by their values in the absence of the field. Taking

$$h^+(t) = h_0\exp(i\omega t + \eta t), \quad (6.107)$$

we find from (6.105) the result, correct to first order in h^+,

$$c^+_{k+q\uparrow}(t)c_{q\downarrow}(t) = \frac{h^+(t)(n_{k+q\uparrow} - n_{q\downarrow})}{(E_q - E_{k+q} - \Delta + \omega - i\eta)} \quad (6.108)$$

The corresponding fluctuation in S^+_k created by the perturbation (6.101) is

$$\langle S^+_k(t)\rangle = \langle \sum_q c^+_{k+q\uparrow}(t)c_{q\downarrow}(t)\rangle, \quad (6.109)$$

and the associated susceptibility $\tilde{\chi}^{(+)}(k,i\omega)$ is defined by the relation

$$\langle S^+_k(t)\rangle = h^+_k(t)\tilde{\chi}^{(+)}(k,i\omega). \quad (6.110)$$

Hence, we obtain the result,

$$\tilde{\chi}^{(+)}(k,i\omega) = P_k(\omega), \quad (6.111)$$

ISBN 0-8053-6610-5, 0-8053-6611-3 (pbk.)

where

$$P_k(\omega) = -\frac{1}{N}\sum_{q} \frac{(f_{q\downarrow} - f_{k+q\uparrow})}{\omega - i\eta + E_q - E_{k+q} - \Delta} \tag{6.112}$$

In (6.112), $f_{k\uparrow}$ is the Fermi distribution function for the state with energy $E_k + U\langle n_\downarrow\rangle$. In keeping with the notation used in other sections, $P_k(\omega)$ is the noninteracting susceptibility for the model.

In order to account for the interactions in the model that give rise to spin waves it is necessary to go back to the stage at which we made the approximation (6.96) for the on-site interaction in the original Hamiltonian. For, in the presence of the perturbation (6.101), terms of the form $\langle c_\sigma^+ c_{-\sigma}\rangle$ are no longer zero as we assumed in making the approximation (6.96). In place of (6.96) we now have for $n_\uparrow n_\downarrow$, say,

$$\begin{aligned} c_\uparrow^+ c_\uparrow c_\downarrow^+ c_\downarrow \simeq \langle n_\downarrow\rangle n_\uparrow + \langle n_\uparrow\rangle n_\downarrow \\ + \langle c_\uparrow c_\downarrow^+\rangle c_\uparrow^+ c_\downarrow + \langle c_\uparrow^+ c_\downarrow\rangle c_\uparrow c_\downarrow^+ . \end{aligned} \tag{6.113}$$

We omit constant terms in (6.113) that do not enter the equation-of-motion for the spin-flip operator. Using the result (6.113) and the corresponding result for $n_\downarrow n_\uparrow$, we replace (6.96) with the approximation,

$$\sum_\sigma n_\sigma n_{-\sigma} \simeq 2\sum_\sigma \langle n_{-\sigma}\rangle n_\sigma + 2\{\langle c_\uparrow^+ c_\downarrow\rangle c_\uparrow c_\downarrow^+ + \langle c_\uparrow c_\downarrow^+\rangle c_\uparrow^+ c_\downarrow\} \tag{6.114}$$

and the approximate Hamiltonian is now

$$\begin{aligned} \mathscr{H} = \sum_{k\sigma} \{E_k + U\langle n_{-\sigma}\rangle\} c_{k\sigma}^+ c_{k\sigma} - U\sum_{\mathbf{l}} \{\langle c_{\mathbf{l}\uparrow}^+ c_{\mathbf{l}\downarrow}\rangle c_{\mathbf{l}\downarrow}^+ c_{\mathbf{l}\uparrow} \\ + \langle c_{\mathbf{l}\downarrow}^+ c_{\mathbf{l}\uparrow}\rangle c_{\mathbf{l}\uparrow}^+ c_{\mathbf{l}\downarrow}\}. \end{aligned} \tag{6.115}$$

We have again ignored the dependence of $\langle n_{\mathbf{l}\sigma}\rangle$ on $\mathbf{l}$ since the first corrections to $\langle n_{\mathbf{l}\sigma}\rangle$ due to the field $h^\nu(t)$ are of second-order in $h^\nu(t)$.

We now repeat the previous calculation of $c_{\mathbf{k+q}\uparrow}^+ c_{\mathbf{q}\downarrow}$ taking the Hamiltonian to be the sum of (6.101) and (6.115). The new term in (6.115) makes the following contribution to the equation-of-motion,

$$-U[c_{\mathbf{k+q}\uparrow}^+ c_{\mathbf{q}\downarrow}, \sum_{\mathbf{l}} \langle c_{\mathbf{l}\uparrow}^+ c_{\mathbf{l}\downarrow}\rangle c_{\mathbf{l}\downarrow}^+ c_{\mathbf{l}\uparrow}] \tag{6.116}$$

and so we need an expression for $\langle c_{\mathbf{l}\uparrow}^+ c_{\mathbf{l}\downarrow}\rangle$. However, we can see from (6.109) and (6.110) that

$$\langle S_{\mathbf{l}}^+\rangle = \langle c_{\mathbf{l}\uparrow}^+ c_{\mathbf{l}\downarrow}\rangle = \frac{1}{N}\sum_{\mathbf{k}} \exp(i\mathbf{k}\cdot\mathbf{l})\tilde{\chi}^{(+)}(k,i\omega)h_{\mathbf{k}}^+(t) \tag{6.117}$$

and this means that the additional term (6.116) gives the following contribution to the equation-of-motion (6.105),

$$-U\tilde{\chi}^{(+)}(k,i\omega)h^+(t)(n_{\mathbf{k+q}\uparrow} - n_{\mathbf{q}\downarrow}).$$

ISBN 0-8053-6610-5, 0-8053-6611-3 (pbk.)

The effect of this latter term on the equation-of-motion is taken into account by making the substitution

$$h^{+}(t) \rightarrow \{1 + U\tilde{\chi}^{(+)}(k,i\omega)\}h^{+}(t),$$

which leads immediately to the final result

$$\tilde{\chi}^{(+)}(k,i\omega) = \{1 + U\tilde{\chi}^{(+)}(k,i\omega)\}P_{k}(\omega)$$

or

$$\tilde{\chi}^{(+)}(k,i\omega) = P_{k}(\omega)\{1 - UP_{k}(\omega)\}^{-1}. \tag{6.118}$$

Notice that (6.118) has exactly the same structure as the RPA result for the electron density susceptibility (6.49), which is not surprising when one reviews the similarities in the steps that lead to the expressions (6.49) and (6.118).

In preparation for the study of the result (6.118) for the transverse spin susceptibility of the Hubbard model we analyze the noninteracting susceptibility $P_{k}(\omega)$, Eq. (6.112). For $k = 0$,

$$\begin{aligned} \text{Im } P_{0}(\omega) &= -\frac{\pi}{N}\sum_{\mathbf{q}} (f_{q\downarrow} - f_{q\uparrow})\,\delta(\omega - \Delta) \\ &= -\pi\delta(\omega - \Delta)(\langle n_{\downarrow}\rangle - \langle n_{\uparrow}\rangle) \\ &= -\pi\frac{\omega}{U}\delta(\omega - \Delta). \end{aligned} \tag{6.119}$$

Hence, the case $k = 0$ merely corresponds to promoting an electron of spin ↓ to the bottom of the spin ↑ band at a cost of energy $\omega = \Delta = U(\langle n_{\downarrow}\rangle - \langle n_{\uparrow}\rangle)$. The spectrum of states for which Im $P_{k}(\omega)$ is nonzero, namely those with an energy $E_{k+q} - E_{q} + \Delta$, is sketched in Fig. (6.1). This spectrum should be compared to the particle-hole spectrum, Fig. (5.2), for nonspin flip excitations. To calculate $P_{k}(\omega)$ for general **k** and ω requires a knowledge of the band energies E_{k}, which in turn requires the calculation of the full band structure [Cooke and Davis (1972) and Cooke (1973)]. An example is shown in Fig. (6.2) which shows the quantity − Im $P_{k}(\omega)$, often called the Stoner density of states, calculated for a realistic band structure of ferromagnetic nickel at absolute zero. A feature to be noted is that for a realistic band structure, where more than one band of electrons is involved, the Stoner density of states for $k \neq 0$ is finite down to $\omega = 0$. This is not readily apparent in Fig. (6.2) because the main structure, which occurs at $\omega \simeq \Delta$, has an amplitude orders of magnitude larger than that at an energy of a few meV.

The poles of the interacting susceptibility (6.118) due to the Hubbard interaction are given by solutions of the equation

$$1 = U \text{ Re } P_{k}(\omega). \tag{6.120}$$

ISBN 0-8053-6610-5, 0-8053-6611-3 (pbk.)

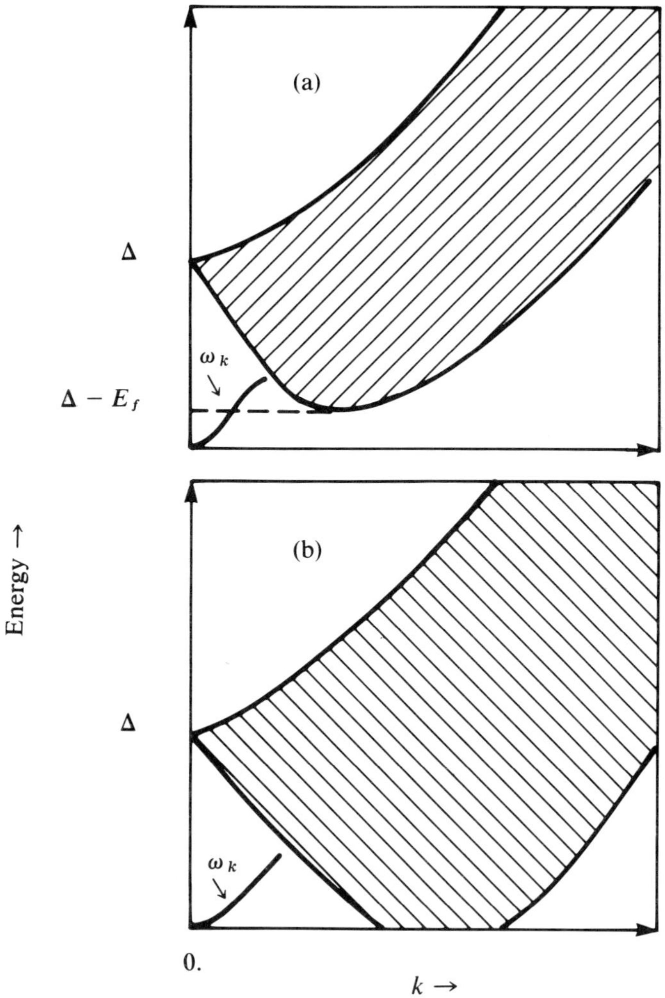

Fig. (6.1) The dissipative part of the noninteracting spin-flip susceptibility, Eq. (6.112), is nonzero within the shaded areas; Fig. (a) is for the case $\Delta > E_f$, and Fig. (b) is for the reverse case, $\Delta < E_f$. The dispersion of the collective mode, or spin wave, ω_k is sketched.

ISBN 0-8053-6610-5, 0-8053-6611-3 (pbk.)

We shall investigate the solution of this equation in the long-wavelength limit. From the definition of $P_k(\omega)$, (6.112)

$$P_k(\omega) = \frac{1}{N}\sum_{\mathrm{q}} f_{q\downarrow}(E_{k+q} + \Delta - E_q - \omega)^{-1} - \frac{1}{N}\sum_{\mathrm{q}} f_{q\uparrow}\,(E_q + \Delta - E_{q-k} - \omega)^{-1}. \quad (6.121)$$

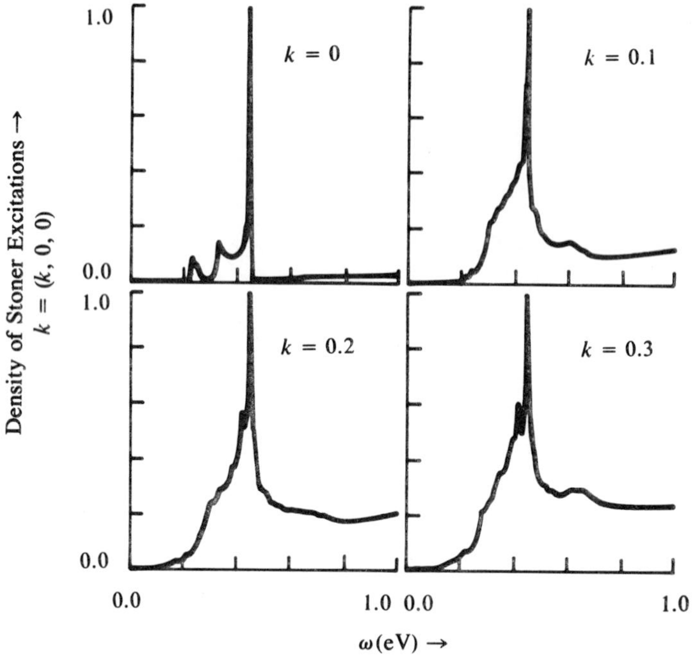

Fig. (6.2) Density of Stoner excitations in ferromagnetic nickel, at absolute zero, for four wavevectors along the (100) Brillouin zone direction. (J. F. Cooke, private communication.)

For small k,

$$E_{k+q} = E_q + \mathbf{k}\cdot\nabla_q E_q + \tfrac{1}{2}(\mathbf{k}\cdot\nabla_q)^2 E_q + \cdots \tag{6.122}$$

and we use this result in (6.121), and expand the denominators in powers of $\mathbf{k}$, assuming that Δ is finite, to obtain, correct to order k^2, the result,

$$\begin{aligned} P_k(\omega) &= \frac{1}{\Delta N}\sum_{\mathbf{q}}(f_{q\downarrow} - f_{q\uparrow}) - \frac{1}{2\Delta^2 N}\sum_{\mathbf{q}}(f_{q\downarrow} + f_{q\uparrow})(\mathbf{k}\cdot\nabla_q)^2 E_q \\ &\quad + \frac{\omega}{\Delta^2 N}\sum_{\mathbf{q}}(f_{q\downarrow} - f_{q\uparrow}) + \frac{1}{\Delta^3 N}\sum_{\mathbf{q}}(f_{q\downarrow} - f_{q\uparrow})(\mathbf{k}\cdot\nabla_q E_q)^2 + \cdots \\ &= \frac{1}{U} - \frac{k^2}{6\Delta^2 N}\sum_{\mathbf{q}}(f_{q\downarrow} + f_{q\uparrow})\nabla_{\mathbf{q}}^2 E_q + \frac{\omega}{U\Delta} \\ &\quad + \frac{k^2}{3\Delta^3 N}\sum_{\mathbf{q}}(f_{q\downarrow} - f_{q\uparrow})(\nabla_{\mathbf{q}} E_q)^2 + \cdots . \end{aligned} \tag{6.123}$$

ISBN 0-8053-6610-5, 0-8053-6611-3 (pbk.)

Substituting the result (6.123) into (6.120) we find the solution

$$\omega_k = k^2 D \tag{6.124}$$

where the spin wave stiffness*

$$D = \frac{U}{6\Delta N}\sum_{\mathrm{q}}(f_{q\uparrow} + f_{q\downarrow})\nabla_{\mathrm{q}}^2 E_q + \frac{U}{3\Delta^2 N}\sum_{\mathrm{q}}(f_{q\uparrow} - f_{q\downarrow})(\nabla_{\mathrm{q}} E)^2. \tag{6.125}$$

We conclude from (6.125) that the transverse susceptibility $\tilde{\chi}^{(+)}(k,i\omega)$ possesses a pole in the long-wavelength limit whose dispersion is proportional to k^2, which is the same as that found for long-wavelength ferromagnetic spin waves in the localized, or Heisenberg model of ferromagnetism. For small k, the collective mode occurs outside the continuum of spin-flip states, as shown in Fig. (6.1), and, in consequence, there is no damping of the mode. However, for larger values of k the mode enters the continuum of spin-flip states and thereby acquires a finite lifetime. This situation is analogous to that of the collective density oscillation, or plasmon, in a degenerate electron gas which has an infinite lifetime for long wavelengths and is damped at short wavelengths through interaction with particle-hole states, cf. Fig. (5.2).

The dissipative part of the interacting susceptibility is from (6.118),

$$\begin{aligned}\operatorname{Im}\tilde{\chi}^{(+)}(k,i\omega) &= \operatorname{Im}\{P_k(\omega)[1 - UP_k(\omega)]^{-1}\}\\ &= \operatorname{Re} P_k(\omega)\operatorname{Im}[1 - UP_k(\omega)]^{-1}\end{aligned} \tag{6.126}$$

where the second equality follows because the frequency ω_k lies outside the continuum of spin-flip states. Now,

$$\operatorname{Im}[1 - UP_k(\omega)]^{-1} = \left(-U\frac{d}{d\omega}P_k\right)_{\omega_k}^{-1}\pi\delta(\omega - \omega_k) \tag{6.127}$$

and evaluating $dP_k/d\omega$, and P_k for $\omega = \omega_k \to 0$ we obtain from (6.126) and (6.127) the result

$$\operatorname{Im}\tilde{\chi}^{(+)}(k,i\omega) = -\pi(\langle n_\downarrow\rangle - \langle n_\uparrow\rangle)\delta(\omega - \omega_k). \tag{6.128}$$

The coefficient of the delta function is proportional to the spin moment per atom and therefore vanishes in the paramagnetic state, as would be expected.

We remind the reader that the calculation given above is for a single electron band. However, essentially the same results are obtained for a

* An exact, finite temperature expression for the spin wave stiffness is obtained by Kishore (1979) using a projection operator formalism similar to that discussed in chapter III.

ISBN 0-8053-6610-5, 0-8053-6611-3 (pbk.)

realistic band structure model of a ferromagnetic metal [Cooke (1973)]. The values of the spin wave stiffness D calculated for iron and nickel are in good agreement with the measured values of 280 meVÅ² and 400 meVÅ², respectively [Cooke and Davis (1972)]. One important difference between our model and a realistic band structure model, which we have already referred to, is that the Stoner density of states is finite for small values of ω in a realistic model. We see from the values of D given for iron and nickel that the spin wave energy for $k = 1\text{Å}^{-1}$, say, is small on the energy scale of the Stoner density of states shown in Fig. (6.2), and that the magnitude of the density of states is relatively small in the spin wave region. Even so, the amplitude of the density of states is sufficient to produce significant damping of the spin wave mode long before its dispersion enters the region where the density of states is significant on the scale on which it is shown in Fig. (6.2). Experiments by Mook *et al.* (1973) and Lynn (1975) show that the spin waves acquire a width comparable to their energy when $\omega_k \sim 150$ meV, and that the magnitude of the damping varies with the direction of the wavevector in the Brillouin zone.

The calculations referred to above are for zero temperature. At finite temperature it is necessary to consider the effects of spin wave damping through interaction with thermally excited spin waves, and electrons excitated out of their zero-temperature configuration. General arguments due to Izuyama and Kubo (1964) show that these two processes lead to terms in the spin wave stiffness that are proportional to $T^{5/2}$ (spin wave-spin wave) and T^2 (spin wave-electron).

ISBN 0-8053-6610-5, 0-8053-6611-3 (pbk.)

BIBLIOGRAPHY

A. A. Abrikosov, L. P. Gorkov, and I. E. Dzyaloshinski. (1963) *Methods of Quantum Field Theory in Statistical Physics* (Englewood Cliffs, New York: Prentice-Hall).

A. A. Abrikosov (1972) *Solid State Physics, Suppl. 12* (New York: Academic Press).

D. J. Amit (1978) *Field Theory, The Renormalization Group and Critical Phenomena* (New York: McGraw-Hill).

T. H. K. Barron and M. L. Klein (1974) *Dynamical Properties of Solids,* Vol. 1, Chap. 5. (ed. G. K. Horton and A. A. Maradudin) (New York: North-Holland).

R. Bausch, H. K. Janssen, and H. Wagner (1976) *Z. Phys. B24,* 113.

N. N. Bogoliubov and D. V. Shirkov (1959) *Introduction to the Theory of Quantized Fields* (New York: Interscience).

J. B. Boyce and B. A. Huberman (1979) Phys. Rep. *51,* 190.

E. Brézin, J. C. Le Guillou, and J. Zinn-Justin (1976) *Phase Transitions and Critical Phenomena,* Vol. 6 (ed. C. Domb and M. S. Green) (London: Academic Press).

C. G. Callan (1970) *Phys. Rev. D2,* 1541.

H. B. Callen (1960) *Thermodynamics* (New York: Wiley).

S. Chandrasekhar (1943) *Rev. Mod. Phys. 15,* 1.

W. Cochran (1969) *Adv. Phys. 18,* 157.

J. C. Collins (1974) *Phys. Rev. D10,* 1213.

J. F. Cooke (1973) *Phys. Rev. B7,* 1108.

J. F. Cooke and H. L. Davis (1972) *AIP Conf. Proc. 10,* 1218.

J. R. D. Copley and S. W. Lovesey (1975) *Rep. Prog. Phys. 38,* 461.

R. A. Cowley and W. J. L. Buyers (1972) *Rev. Mod. Phys. 44,* 406.

C. De Dominicis and L. Peliti (1978) *Phys. Rev. B18,* 353.

U. Deker and F. Haake (1975a) *Phys. Rev. A11,* 2043; (1975b) *ibid. A12,* 1629.

C. Di Castro and G. Jona-Lasinio (1976) *Phase Transitions and Critical Phenomena,* Vol. 6 (ed. C. Domb and M. S. Green) (London: Academic Press).

O. W. Dietrich, J. Als-Nielsen, and L. Passell (1976) *Phys. Rev. B14,* 4932.

S. Doniach and E. H. Sondheimer (1974) *Green's Functions for Solid State Physicists* (Reading, Massachusetts: Benjamin, Advanced Book Program).

F. J. Dyson (1956) *Phys. Rev. 102,* 1217; *ibid.* 1230.

A. Einstein (1910) *Ann. Phys. 33,* 1275.

P. Eisenberger, P. M. Platzman, and P. Schmidt (1975) *Phys. Rev. Lett. 34,* 18.

R. J. Elliott, J. A. Krumhansl, and P. L. Leath (1974) *Rev. Mod. Phys. 46,* 465.

R. Englman and P. Levy (1963) *J. Math. Phys. 4,* 105.

M. H. Ernst and J. R. Dorfman (1975) *J. Stat. Phys. 12,* 311.

ISBN 0-8053-6610-5, 0-8053-6611-3 (pbk.)

A. L. Fetter and J. D. Walecka (1971) *Quantum Theory of Many-Particle Systems* (New York: McGraw-Hill).
D. Forster (1975) *Hydrodynamic Fluctuations, Broken Symmetry, and Correlation Functions* (Reading, Massachusetts: Benjamin, Advanced Book Program).
D. Forster, D. R. Nelson, and M. J. Stephen (1977) *Phys. Rev. A16,* 732.
R. F. Fox (1978) *Phys. Rep. 48,* 180.
L. Garrido and M. San Miguel (1978) *Prog. Theor. Phys. 59,* 40; *ibid. 59,* 55.
M. Gitterman (1978) *Rev. Mod. Phys. 50,* 85.
R. Graham (1973) *Springer Tracts in Modern Physics, 66* (Berlin: Springer-Verlag).
M. S. Green (1952) *J. Chem. Phys. 20,* 1281.
M. P. Greene, H. J. Lee, J. J. Quinn, and S. Rodriguez (1969) *Phys. Rev. 177,* 1019.

J. P. Hansen and I. R. McDonald (1976) *Theory of Simple Liquids* (New York: Academic Press).
A. B. Harris (1968) *Phys. Rev. 175,* 674; *ibid.* (1969) *184,* 606.
C. Herring (1966) *Magnetism,* Vol. IV (ed. G. T. Rado and H. Suhl) (New York: Academic Press).
P. C. Hohenberg and B. I. Halperin (1977) *Rev. Mod. Phys. 49,* 1.
J. Hubbard (1963) *Proc. Roy. Soc. A 276,* 238.
J. Hubbard (1979) *Phys. Rev. B19,* 2626.
J. Hubbard and J. L. Beeby (1969) *J. Phys. C2,* 556.

T. Izuyama and R. Kubo (1964) *J. Appl. Phys. 35,* 1074.

L. P. Kadanoff and G. Baym (1962) *Quantum Statistical Mechanics* (Reading, Massachusetts: Benjamin, Advanced Book Program).
T. Karasudani, N. Nagano, H. Okamoto, and H. Mori (1969) *Prog. Theor. Phys. 61,* 850.
K. Kawasaki (1976) *Phase Transitions and Critical Phenomena,* Vol. 5A (eds. C. Domb and M. S. Green) (New York: Academic Press).
F. Keffer (1967) *Encyclopedia of Physics,* Vol. XVIII/2 (Berlin: Springer-Verlag).
R. Kishore (1979) *Phys. Rev. B19,* 3822.
W. Kohn (1959) *Phys. Rev. Lett. 2,* 393.
R. H. Kraichnan (1972) *Proc. 6th IUPAP Conf. Statistical Mechanics,* (eds. S. A. Rice, K. F. Freed, and J. C. Light) (Chicago: University of Chicago).
R. Kubo (1957) *J. Phys. Soc. Japan 12,* 570.
R. Kubo (1966) *Rep. Prog. Phys. 29,* 255.

H. Lamb (1932) *Hydrodynamics* (London: Cambridge University Press).
L. D. Landau (1930) *Z. Phys. 64,* 629.
L. D. Landau and E. M. Lifshitz (1959) *Statistical Physics* (London: Pergamon Press).
L. D. Landau and E. M. Lifshitz (1963) *Fluid Mechanics* (Oxford: Pergamon Press).
L. D. Landau and E. M. Lifshitz (1965) *Quantum Mechanics* (Oxford: Pergamon Press).
D. Levesque and L. Verlet (1970) *Phys. Rev. A2,* 2514.
J. W. E. Lewis and S. W. Lovesey (1978) *J. Phys. C11,* L57.
J. M. Loveluck and E. Balcar (1979) *Phy. Rev. Let 42,* 1563.
J. M. Loveluck and C. G. Windsor (1978) *J. Phys. C11,* 2999.
S. W. Lovesey and J. M. Loveluck (1976) *J. Phys. C9,* 3639; *ibid.,* 3659.
J. W. Lynn (1975) *Phys. Rev. B11,* 2624.

ISBN 0-8053-6610-5, 0-8053-6611-3 (pbk.)

S. K. Ma and G. F. Mazenko (1975) *Phys. Rev. B11,* 4077.
S. K. Ma (1976) *Modern Theory of Critical Phenomena* (Reading, Massachusetts: Benjamin, Advanced Book Program).
D. K. C. MacDonald (1962) *Noise and Fluctuations* (New York: Wiley).
A. R. Mackintosh and H. B. Møller (1972) *Magnetic Properties of Rare Earth Metals* (ed. R. J. Elliott) (New York: Plenum Press).
A. A. Maradudin, E. W. Montroll, and G. M. Weiss (1963) *Solid State Physics Suppl. 3* (New York: Academic Press).
P. C. Martin, E. D. Siggia, and H. A. Rose (1973) *Phys. Rev. A8,* 423.
D. C. Mattis (1965) *The Theory of Magnetism* (New York: Harper and Row).
A. S. Monin and A. M. Yaglom (1975) *Statistical Fluid Mechanics,* Vol. 2 (Cambridge, Massachusetts: MIT Press).
H. A. Mook, J. W. Lynn, and R. M. Nicklow (1973) *Phys. Rev. Lett. 30,* 556.
H. Mori (1965a) *Prog. Theor. Phys. 33,* 423; (1965b) *Prog. Theor. Phys. 34,* 399.
H. Mori and H. Fujisaka (1973) *Prog. Theor. Phys. 49,* 764.

C. Nash (1978) Relativistic Quantum Fields (London: Academic Press).
G. Nicolis (1979) Rep. Prog. Phys. *42,* 226.
B. R. A. Nijboer and A. Rahman (1966) *Physica 32,* 415.
E. A. Novikov (1965) *JETP Lett. 20,* 1290.

R. E. Peierls (1955) *Quantum Theory of Solids* (Oxford: Oxford University Press).
D. Pines and P. Nozières (1966) *The Theory of Quantum Liquids* (New York: Benjamin).
P. M. Platzman and P. A. Wolff (1973) *Solid State Physics Suppl. 13* (New York: Academic Press).
P. M. Platzman and P. Eisenberger (1974) *Phys. Rev. Lett. 33,* 152.
Y. Pomeau and P. Résibois (1975) *Phys. Rep. 19,* 63.

E. Rastelli and P-A Lindgård (1979) *J. Phys. (C) 12,* 1899.
R. D. Reed and R. R. Roy (1971) *Statistical Physics for Students of Science and Engineering* (Scranton: Intext Educational Publishers).
P. Résibois and M. de Leener (1977) *Classical Kinetic Theory of Fluids* (New York: Wiley).
S. M. Rezende and R. M. White (1978) *Phys. Rev. B18,* 2346.

G. A. Samara (1977) *Comments Sol. State Phys. 8,* 13.
M. Sargent, M. O. Scully, and W. E. Lamb (1974) *Laser Physics* (Reading, Massachusetts: Addison-Wesley, Advanced Book Program).
L. I. Schiff (1955) *Quantum Mechanics* (New York: McGraw-Hill).
T. Schneider and E. Stoll (1978) *Phys. Rev. B17,* 1302.
P. Schofield (1975) *Specialist Reps., Stat. Mech.* II (London: Chemical Society).
K. D. Schotte (1978) *Phys. Rep. 46,* 94.
G. L. Squires (1978) Introduction to the Theory of Thermal Neutron Scattering (Cambridge: Cambridge University Press)
H. E. Stanley (1971) *Introduction to Phase Transitions and Critical Phenomena* (Oxford: Oxford University Press).
M. Steiner, J. Villain, and C. G. Windsor (1976) *Adv. Phys. 25,* 87.
H. L. Swinney and D. L. Henry (1973) *Phys. Rev. A8,* 2586.
H. L. Swinney and J. P. Gollub (1978) *Physics Today,* August, 41.
K. Symanzik (1970) *Commun. Math. Phys. 18,* 227; (1971) *Ibid. 23,* 49.

ISBN 0-8053-6610-5, 0-8053-6611-3 (pbk.)

J. C. Taylor (1976) *Gauge Theories of Weak Interactions* (Cambridge: Cambridge University Press).
G. 't Hooft (1973) *Nucl. Phys. B61,* 455.

L. Van Hove (1954) *Phys. Rev. 95,* 249.

M. Wadati (1979) *Phys. Rep. 50,* 88.
N. Wakabayashi, R. M. Nicklow, and H. G. Smith (1971) *Phys. Rev. B4,* 2558.
D. J. Wallace and R. K. P. Zia (1978) *Rep. Prog. Phys. 41,* 1.
S. Weinberg (1977) *Phys. Today,* April, 42.
K. G. Wilson and J. B. Kogut (1974) *Phys. Rep. 12,* 75.

R. Zeyher (1978) *Z. Phys. B31,* 127.
J. M. Ziman (1964) *Principles of the Theory of Solids* (Cambridge: Cambridge University Press).
J. M. Ziman (1969) *Elements of Advanced Quantum Theory* (Cambridge: Cambridge University Press).
D. N. Zubarev (1960) *Sov. Phys. Usp. 3,* 320.
R. Zwanzig (1972) *Proc. 6th IUPAP Conf. Statistical Mechanics,* (eds. S. A. Rice, K. F. Freed, and J. C. Light) (Chicago: University of Chicago Press).

ISBN 0-8053-6610-5, 0-8053-6611-3 (pbk.)

INDEX